New Insights in Stability, Structure and Properties of Porous Materials

Special Issue Editors

Annalisa Martucci
Giuseppe Cruciani

MDPI • Basel • Beijing • Wuhan • Barcelona • Belgrade

MDPI

Special Issue Editors
Annalisa Martucci
University of Ferrara
Italy

Giuseppe Cruciani
University of Ferrara
Italy

Editorial Office
MDPI AG
St. Alban-Anlage 66
Basel, Switzerland

This edition is a reprint of the Special Issue published online in the open access journal *Minerals* (ISSN 2075-163X) in 2017 (available at: http://www.mdpi.com/journal/minerals/special_issues/Porous_Materials).

For citation purposes, cite each article independently as indicated on the article page online and as indicated below:

Author 1; Author 2. Article title. *Journal Name*. **Year**. *Article number*, page range.

First Edition 2017

ISBN 978-3-03842-450-5 (Pbk)
ISBN 978-3-03842-451-2 (PDF)

Table of Contents

About the Special Issue Editors

Annalisa Martucci is an Associate Professor of Applied Mineralogy at the Department of Earth Sciences of the University of Ferrara (Italy). She graduated in Geological Sciences in 1995 at the University of Bari, Italy, in 1999 received a Ph.D. in Mineralogy and Crystallography at the University of Ferrara, Italy. Her research interests focus on the crystallography and crystal-chemistry of heterogeneous catalysts, in particular zeolites and related porous solids. She is an expert on single crystal and powder diffraction, both with conventional X-rays and large scale facility radiations (synchrotron X-rays and neutrons). She served as a council member of the Italian Zeolite Association and has authored more than 60 scientific papers on international refereed journals in the fields of mineralogy, applied mineralogy and solid state chemistry.

Giuseppe Cruciani is Full Professor of Mineralogy at the University of Ferrara. He graduated with honor in Geological Sciences in 1989 at the University of Perugia, where he received a PhD in Mineralogy (Crystallography) and Petrology in 1993. He was a visiting scientist in 1995 at the European Synchrotron Radiation Facility (ESRF) in Grenoble, France. From 2008–2010 he was a member of the chemistry review committee for project selection at ESRF. Since 2016 he has held the position of Chair of the proposal review panel for hard condensed matter-structures at the Elettra synchrotron in Trieste, Italy. From 2012–2013 he was President of the Italian Society of Mineralogy and Petrology and is currently (2016–2019) President of the Italian Zeolite Association. He serves as Associate Editor of the European Journal of Mineralogy. His main research fields cover crystallography and crystal-chemistry of zeolite-like minerals and their synthetic analogues, and of many other silicate and oxide systems. Experimental projects have mostly focused on single crystal and powder diffraction, both with conventional X-rays and large scale facility radiation (synchrotron X-rays and neutrons). He has authored more than 130 articles in international journals and more than 10 book chapters, and his H-index is equal to 27 (as of May 2017).

Preface to "New Insights in Stability, Structure and Properties of Porous Materials"

The papers collated here are a selection of papers concerning new insights into porous materials and their applications in many technological and environmental fields, such as catalysis, adsorption, separation and ion exchange.

Following the overall thread governing this series in Minerals, this Special Issue aimed to be a bridge to bring experimental and theoretical scientists together, with the aim of exchanging information and discussing recent developments regarding research on porous materials. Moreover, it was intended to emphasize the relationships between the structure and/or chemical composition and the specific physical properties of these materials, their role in mineralogical, technological, green as well as sustainable processes. The selected high-quality original and review papers concern the physical, chemical and structural characterization of porous materials, synthesis of crystalline phases with pores in the appropriate range, structure–property relationships at ambient conditions, but also at low and high temperatures and/or at high pressures, adsorption and diffusion of mobile species in porous materials, host–guest interactions and confinement effects, ion exchange, modeling in geological and environmental processes, new insights in processing and applications.

As can be clearly seen, the particularity if this Special Issue is its interdisciplinary character.

It is our hope that this collection will serve as a valuable and substantive resource for anyone interested in studies of porous materials, as well as satisfy the curiosity of readers and encourage others to further pursue their interest in relationships between the structure and/or chemical composition and the specific physical properties of these materials, and their role in mineralogical, technological, green and sustainable processes.

Before closing, we would like to acknowledge all the authors, the scientific board in editing this Special Issue who we owe many thanks, as well as the review board.

Annalisa Martucci and Giuseppe Cruciani
Special Issue Editors

Editorial

Editorial for Special Issue "New Insights in Stability, Structure and Properties of Porous Materials"

Annalisa Martucci * and Giuseppe Cruciani

Department of Physics and Earth Sciences, University of Ferrara, Via Saragat 1, 44122 Ferrara, Italy; giuseppe.cruciani@unife.it
* Correspondence: mrs@unife.it; Tel.: +39-0532-974730

Academic Editor: Paul Sylvester
Received: 5 May 2017; Accepted: 8 May 2017; Published: 11 May 2017

Porous materials (such as zeolites, clay minerals, and assemblies of oxide nanoparticles) are of great importance for the progress in many technological and environmental fields, such as catalysis, adsorption, separation, and ion exchange, because of their unique pore topologies, tunable structures, and the possibility of introducing active reaction sites.

The major goal of this special issue is to provide a platform for scientists to discuss new insights in the stability, structure, and properties of porous materials, as well as in innovative aspects in their processing and applications. The emphasis is on the relationships between the structure and/or chemical composition and the specific physical properties of these materials, as well as their role in mineralogical, technological, green, and sustainable processes. With this special issue of *Minerals*, we have endeavored to provide an up-to-date selection of high-quality original and review papers concerning the physical, chemical, and structural characterization of porous materials, the synthesis of crystalline phases with pores in the appropriate range, structure–property relationships at ambient conditions but also at high temperatures and/or at high pressures, adsorption, and diffusion of mobile species in porous materials, host/guest interactions and confinement effects, ion exchange, modeling in geological and environmental processes, and new insights in processing and applications. In total, eight fashionable contributions reflect both the diversity and interdisciplinary of modern mineralogy, bridging together experimentalists and computational approaches.

The review presented by Bandura et al. [1] is dedicated to the decontamination strategies available today for the removal of petroleum substances and their derivatives from roads, water, and air. Specifically, this paper presents an overview of recent research papers concerning porous (natural, synthetic, and modified mineral adsorbents) materials used as adsorbents for petroleum pollutants, present in water and spilled on land, occurring as oils, petroleum industry derivatives, and volatile compounds. Environmental pollution with petroleum products has become a major problem worldwide and is a consequence of industrial growth. The development of sustainable methods for the removal of petroleum substances and their derivatives from aquatic and terrestrial environments and from air has therefore become extremely important today.

Advanced technologies and materials dedicated to this purpose are relatively expensive. Among several techniques developed for BTEX (benzene, toluene, ethylbenzene, and xylene) removal from waters, adsorption is one of the most efficient methods, thanks to satisfactory efficiencies (even at low concentrations), easy operation, and low cost [2,3].

Recently, adsorption on hydrophobic zeolites has received the greatest interest in water treatment technology due to their organic contaminant selectivity, thermal and chemical stability, strong mechanical properties, rapid kinetics, and absence of salt and humic substance interference [4–10]. In this issue, the Sarti et al. [11] contribution is dedicated to this topic and is focused on the adsorption of toluene from aqueous solutions onto hydrophobic beta zeolites by combining chromatographic, thermal, and structural techniques. This work highlights the differences in adsorption properties between as-synthesized and calcined beta zeolites, with different SiO_2/Al_2O_3 ratios, toward a water contaminant of great concern such as toluene. The authors demonstrate that the thermal treatment significantly improves the adsorption properties of all selected zeolites especially for the most hydrophobic beta, thus opening new alternatives for the industrial application of this material, mainly in hydrocarbon adsorption processes in the presence of water.

In order for the adsorption process to be cost-effective, the progressive deactivation of saturated sorbents has become an essential task [12–16]. Thermal treatment is the most common regeneration technique, where organic host molecules are decomposed and/or oxidized at high temperature. Consequently, there is a strong interest in understanding the mechanisms behind the thermal regenerative solution, which makes zeolites regenerable materials that are efficiently reusable in the contaminant adsorption process. In this issue, the temperature-induced desorption of methyl *tert*-butyl ether (MTBE) from aqueous solutions onto hydrophobic ZSM-5 zeolite is studied by Rodeghero et al. [17] using in situ synchrotron powder diffraction and chromatographic techniques. Rietveld analysis demonstrated that the desorption process occurred without any significant zeolite crystallinity loss, but with slight deformations in the channel apertures. This kind of information is crucial for understanding the features of both adsorption and desorption processes, thus helping in the design of water treatment appliances based on microporous materials as well as designing and optimizing the regeneration treatment of zeolite.

As reported in this issue by Bundru et al. [18], the regeneration of the exhausted zeolite as well as the recovery of ammonia are feasible processes. Spent exchangers such as NH_4-exchanged synthetic zeolites can be transformed into mullite and amorphous silica by thermal treatments [19–21]. With this perspective, a material containing NH_4-clinoptilolite, derived from a wastewater treatment, has be evaluated as a potential raw material for the ceramic industry. The results of this research are interesting, because they indicate that NH_4-clinoptilolite represents a raw material of interest in the ceramic field, in particular in the production of acid refractory.

The reuse (addition) of the spent zeolitic sorbents containing petroleum waste to produce lightweight aggregates (LWAs) is also discussed by Franus et al. [22]. It is well known that the mineral composition and organic amendments to the substrate can control the physical properties of LWAs. Therefore, Franus et al. [22] hypothesize here that the addition of waste zeolites can modify the structure of the standard clay-based LWAs towards higher porosity, which differs depending on the zeolite used.

As reported by Arletti et al. [23], recent studies on the behavior of both natural and synthetic microporous materials under high pressure (HP) provide important information on their elastic behavior and stability, thus opening new perspectives for technological applications. This paper presents a study, performed by in situ synchrotron X-ray powder diffraction (XRPD), of the HP stability and behavior of the natural zeolite amicite. The investigation aimed in particular to understand the relationships between compressibility and framework/extraframework content as well as the influence of different penetrating or non-penetrating fluids on the compressibility and HP deformation mechanisms of this zeolite.

In the present volume, Krupskaya et al. [24] discuss the mechanism of montmorillonite structural alteration and the modification of bentonites' properties under thermochemical treatment (treatment with inorganic acid solutions at different temperatures, concentrations, and reaction times). The mechanism of montmorillonite transformation under acid solution treatment as well as its influence on bentonite properties are evaluated. The modification of structural and adsorption characteristics with acid treatment can be useful to simulate behavior of the engineered barrier properties for repositories of radioactive and industrial wastes, especially in the case of dealing with liquid radioactive wastes.

The aim of the Steudel et al. [25] study is the characterization of a clay from the Madrid basin, which shows exceptional suitability as adsorbent material in biotechnology processes [26], as adsorbent for mycotoxins [27] as well as in pesticide removal from water [28] for this clay. This last can be also used to bind contaminants from the manufacture of paper [29]. The authors reported that this clay is highly suitable for mining without chemical pretreatment, which reduces environmental burden [29].

Finally, the Due et al. study [30] is focused on volumetric swelling strain and strength reduction of pillars when CO_2 is stored in abandoned coal mines. The volumetric swelling strain is theoretically derived as a function of time by adsorption pressure increasing step by step under unconfined conditions. In connection with the conditions of coal pillars in abandoned coal mines, and a uniaxial loading model is proposed by simplifying the actual condition.

In conclusion, it is my hope that this special issue will serve as a valuable and substantive resource for anyone interested in studies of porous materials, as well as satisfy the curiosity of readers and encourage others to pursue further their interest in relationships between the structure and/or chemical composition and the specific physical properties of these materials, as well as their role in mineralogical, technological, green, and sustainable processes.

Conflicts of Interest: The authors declare no conflict of interest.

References

1. Bandura, L.; Woszuk, A.; Kołodyńska, D.; Franus, W. Application of Mineral Sorbents for Removal of Petroleum Substances: A Review. *Minerals* **2017**, *7*, 37. [CrossRef]
2. Gupta, V.K.; Verma, N. Removal of volatile organic compounds by cryogenic condensation followed by adsorption. *Chem. Eng. Sci.* **2002**, *57*, 2679–2696. [CrossRef]
3. Pasti, L.; Rodeghero, E.; Sarti, E.; Bosi, V.; Cavazzini, A.; Bagatin, R.; Martucci, A. Competitive adsorption of VOCs from binary aqueous mixtures on zeolite ZSM-5. *RSC Adv.* **2016**, *6*, 54544–54552.
4. Costa, A.A.; Wilson, W.B.; Wang, H.; Campiglia, A.D.; Dias, J.A.; Dias, S.C.L. Comparison of BEA, USY and ZSM-5 for the quantitative extraction of polycyclic aromatic hydrocarbons from water samples. *Microporous Mesoporous Mater.* **2012**, *149*, 186–192. [CrossRef]
5. Abu-Lail, L.; Bergendahl, J.A.; Thompson, R.W. Adsorption of methyl tertiary butyl ether on granular zeolites: Batch and column studies. *J. Hazard. Mater.* **2010**, *178*, 363–369. [CrossRef] [PubMed]
6. Anderson, M.A. Removal of MTBE and other organic contaminants from water by sorption to high silica zeolites. *Environ. Sci. Technol.* **2000**, *34*, 725–727. [CrossRef]
7. Rossner, A.; Knappe, D.R. MTBE adsorption on alternative adsorbents and packed bed adsorber performance. *Water Res.* **2008**, *42*, 2287–2299. [CrossRef] [PubMed]
8. Pasti, L.; Martucci, A.; Nassi, M.; Cavazzini, A.; Alberti, A.; Bagatin, R. The role of water in DCE adsorption from aqueous solutions onto hydrophobic zeolites. *Microporous Mesoporous Mater.* **2012**, *160*, 182–193. [CrossRef]
9. Pasti, L.; Sarti, E.; Cavazzini, A.; Marchetti, N.; Dondi, F.; Martucci, A. Factors affecting drug adsorption on beta zeolites. *J. Sep. Sci.* **2013**, *36*, 1604–1611. [CrossRef] [PubMed]
10. Braschi, I.; Martucci, A.; Blasioli, S.; Mzini, L.L.; Ciavatta, C.; Cossi, M. Effect of humic monomers on the adsorption of sulfamethoxazole sulfonamide antibiotic into a high silica zeolite Y: An interdisciplinary study. *Chemosphere* **2016**, *155*, 444–452. [CrossRef] [PubMed]
11. Sarti, E.; Chenet, T.; Pasti, L.; Cavazzini, A.; Rodeghero, E.; Martucci, A. Effect of Silica Alumina Ratio and Thermal Treatment of Beta Zeolites on the Adsorption of Toluene from Aqueous Solutions. *Minerals* **2017**, *7*, 22. [CrossRef]
12. Leardini, L.; Martucci, A.; Braschi, I.; Blasioli, S.; Quartieri, S. Regeneration of high-silica zeolites after sulfamethoxazole antibiotic adsorption: A combined in situ high-temperature synchrotron X-ray powder diffraction and thermal degradation study. *Mineral. Mag.* **2014**, *78*, 1141–1160. [CrossRef]
13. Martucci, A.; Rodeghero, E.; Pasti, L.; Bosi, V.; Cruciani, G. Adsorption of 1,2-dichloroethane on ZSM-5 and desorption dynamics by in situ synchrotron powder X-ray diffraction. *Microporous Mesoporous Mater.* **2015**, *215*, 175–182. [CrossRef]
14. Braschi, I.; Blasioli, S.; Buscaroli, E.; Montecchio, D.; Martucci, A. Physicochemical regeneration of high silica zeolite Y used to clean-up water polluted with sulfonamide antibiotics. *J. Environ. Sci.* **2016**, *43*, 302–312. [CrossRef] [PubMed]
15. Guisnet, M.; Ribeiro, F.R. *Deactivation and Regeneration of Zeolite Catalysts*; World Scientific: Singapore, 2011.

16. Wu, Z.; An, Y.; Wang, Z.; Yang, S.; Chen, H.; Zhou, Z.; Mai, S. Study on zeolite enhanced contact-dsorption regeneration-stabilization process for nitrogen removal. *J. Hazard. Mater.* **2008**, *156*, 317–326. [CrossRef] [PubMed]
17. Rodeghero, E.; Pasti, L.; Sarti, E.; Cruciani, G.; Bagatin, R.; Martucci, A. Temperature-Induced Desorption of Methyl *tert*-Butyl Ether Confined on ZSM-5: An In Situ Synchrotron XRD Powder Diffraction Study. *Minerals* **2017**, *7*, 34. [CrossRef]
18. Brundu, A.; Cerri, G.; Sale, E. Thermal Transformation of NH_4-Clinoptilolite to Mullite and Silica Polymorphs. *Minerals* **2017**, *7*, 11. [CrossRef]
19. Matsumoto, T.; Goto, Y.; Urabe, K. Formation process of mullite from NH_4^+-exchanged Zeolite A. *J. Ceram. Soc. Jpn.* **1995**, *103*, 93–95. [CrossRef]
20. Kosanović, C.; Subotić, B.; Smit, I. Thermally induced phase transformations in cation-exchanged zeolites 4A, 13X and synthetic mordenite and their amorphous derivatives obtained by mechanochemical treatment. *Thermochim. Acta* **1998**, *317*, 25–37. [CrossRef]
21. Kosanović, C.; Subotić, B. Preparation of mullite micro-vessels by a combined treatment of zeolite A. *Microporous Mesoporous Mater.* **2003**, *66*, 311–319. [CrossRef]
22. Franus, W.; Jozefaciuk, G.; Bandura, L.; Franus, M. Use of Spent Zeolite Sorbents for the Preparation of Lightweight Aggregates Differing in Microstructure. *Minerals* **2017**, *7*, 25. [CrossRef]
23. Arletti, R.; Giacobbe, C.; Quartieri, S.; Vezzalini, G. The Influence of the Framework and Extraframework Content on the High Pressure Behavior of the GIS Type Zeolites: The Case of Amicite. *Minerals* **2017**, *7*, 18. [CrossRef]
24. Krupskaya, V.; Zakusin, S.; Tyupina, E.; Dorzhieva, O.; Zhukhlistov, A.; Belousov, P.; Timofeeva, M. Experimental Study of Montmorillonite Structure and Transformation of Its Properties under Treatment with Inorganic Acid Solutions. *Minerals* **2017**, *7*, 49. [CrossRef]
25. Steudel, A.; Friedrich, F.; Schuhmann, R.; Ruf, F.; Sohling, U.; Emmerich, K. Characterization of a Fine-Grained Interstratification of Turbostratic Talc and Saponite. *Minerals* **2017**, *7*, 5. [CrossRef]
26. Temme, H.; Sohling, U.; Suck, K.; Ruf, F.; Niemeyer, B. Separation of aromatic alcohols and aromatic ketones by selective adsorption on kerolite-stevensite clay. *Colloids Surf. A* **2011**, *377*, 290–296. [CrossRef]
27. Sohling, U.; Haimerl, A. Use of Stevensite for Mycotoxin Adsorption. Patent WO 2006119967, 17 July 2012.
28. Ureña-Amate, M.D.; Socías-Viciana, M.; González-Pradas, E.; Saifi, M. Effects of ionic strength and temperature on adsorption of atrazine by a heat treated kerolite. *Chemosphere* **2005**, *59*, 69–74. [CrossRef] [PubMed]
29. Sohling, U.; Ruf, F. Stevensite and/or Kerolite Containing Adsorbents for Binding Interfering Substances during the Manufacture of Paper. Patent WO 200702941, 1 March 2007.
30. Du, Q.; Liu, X.; Wang, E.; Wang, S. Strength Reduction of Coal Pillar after CO_2 Sequestration in Abandoned Coal Mines. *Minerals* **2017**, *7*, 26. [CrossRef]

minerals MDPI

Article

Characterization of a Fine-Grained Interstratification of Turbostratic Talc and Saponite

Annett Steudel [1,*], **Frank Friedrich** [2], **Rainer Schuhmann** [1], **Friedrich Ruf** [3], **Ulrich Sohling** [3,4] **and Katja Emmerich** [1]

[1] Competence Center for Material Moisture (CMM), Karlsruhe Institute of Technology, Hermann-von-Helmholtz-Platz 1, D-76344 Eggenstein-Leopoldshafen, Germany; rainer.schuhmann@kit.edu (R.S.); katja.emmerich@kit.edu (K.E.)

[2] Chair of Foundation Engineering, Soil- and Rock Mechanics, Ruhr-University Bochum, Universitätsstraße 150, 44780 Bochum, Germany; fwfriedrich@googlemail.com

[3] Clariant Produkte (Deutschland) GmbH, BU Functional Minerals, BL Adsorbents, Ostenrieder Str. 15, 85368 Moosburg, Germany; friedrich.ruf@clariant.com (F.R.); ulrich.sohling@clariant.com (U.S.)

[4] Clariant Produkte (Deutschland) GmbH, Competence Center Colorants & Functional Chemicals, Group Technology & Innovation, Industriepark Höchst, Gebäude G 860 (CIC), 65926 Frankfurt, Germany

* Correspondence: annett.steudel@kit.edu; Tel.: +49-721-608-26805; Fax: +49-721-608-23478

Academic Editors: Annalisa Martucci and Huifang Xu
Received: 8 November 2016; Accepted: 23 December 2016; Published: 5 January 2017

Abstract: Interstratifications of talc and trioctahedral smectites from different provenances are used as indicators for geological environments and for geotechnical and technical applications. However, comprehensive layer characterization of these interstratifications is rare. Sample EX M 1694, a clay with red-beige appearance from the Madrid basin was studied by X-ray diffraction analysis, X-ray fluorescence analysis, Fourier transformation infrared spectroscopy, simultaneous thermal analysis, gas adsorption measurements, cation exchange capacity, and environmental scanning electron microscopy. More than 95% of particles in EX M 1964 belong to the clay fraction <2 µm. It contains 75% interstratification of 30% turbostratic talc, and 70% saponite type III and 25% turbostratic talc. The turbostratic talc(0.3)/saponite interstratification is characterized by a low number of layers per stack (3), small lateral dimension of layers (60–80 nm) and, accordingly, a high specific surface area (283 m^2/g) with nearly equal surface area of micro- and mesopores. Thus, the studied material can be used as mined for adsorption, in contrast to acid-treated clays that produce hazardous waste during production. Low particle size of the interstratification drastically reduced thermal stability and dehydroxylation was superimposed by recrystallization of high temperature phases already at 816 °C, which is low for trioctahedral 2:1 layer minerals.

Keywords: talc; kerolite; saponite; stevensite; mixed layer; modelling of one-dimensional X-ray pattern; simultaneous thermal analysis

1. Introduction

Interstratifications of talc and trioctahedral smectite layers are formed as an abundant mineral in lake and/or spring deposits of Miocene to Pleistocene age and in serpentinized rocks formed by a transformation of ultramafic rocks at low temperature.

Turbostratic talc/trioctahedral smectite interstratifications occur, for example, in the Province Parma (Italy) [1], in the Armagosa Desert (Nevada) [2], and in the Madrid basin (Spain) [3–6], which is the most extensive studied locality. The interstratifications are often very fine-grained and analysis and description of the interstratification is still difficult. Turbostratic talc has the same chemical composition as talc, but displays a fully turbostratic structure [7–9] and was called kerolite or disordered talc in the past. The term kerolite was discredited by the International Mineralogical Association (IMA)/Commission on new minerals, nomenclature, and classification (CNMNC) in 2008 [10,11], but is still in use in recent literature e.g., Guggenheim [12], because turbostratic talc is regarded as a variety of talc. Nevertheless, in this paper turbostratic talc will be used according to IMA regulations. Stacks of turbostratic talc contain less than 4–5 layers and show broad basal reflections with an increased basal spacing of about 0.96 ± 0.005 nm compared to talc (0.936 nm; [13]). In the talc structure the oxygen atoms of adjacent layers are partially packed together, thereby, the layers are close together. If the distance between the oxygen atoms changed with rotation of layers, the layers would be 0.027 nm further apart. Thus, the sum of the basal reflection of talc 0.936 nm plus 0.027–0.029 nm due to disorder, gives 0.963–0.965 nm, which corresponds with the observed spacing of turbostratic talc [7]. Specific surface area, determined by gas adsorption methods, is about 200 m^2/g, which also indicates small particle size. Compared to talc, turbostratic talc is supposed to hold additional water in the structure, which is probably mainly surface-held water [7].

One challenge is to identify the character of the trioctahedral smectite (saponite vs. stevensite) in the interstratifications. Stevensite and saponite are hydrous magnesium silicates belonging to the smectite group [14] with a 060 reflection at 0.152 nm. Stevensite differs from saponite by a complete absence of trivalent cations (Al and Fe(III)). The resulting layer charge of stevensite is caused by a deficiency of octahedral cations. The layer charge of stevensite is at the lower limit (near 0.2 per formula unit, p.f.u.) for known layer charges of smectite [15,16]. In contrast, saponite shows a larger variability in chemical composition. Saponites are characterized by a higher layer charge p.f.u. (0.3–0.5) that is more common for smectites. Saponite type I is characterized by tetrahedral substitutions without octahedral substitutions [16,17], whereas saponite type II is characterized by an additional positive octahedral charge [18]. Furthermore, saponites may be either iron-free or have octahedral iron, which results in saponite type III [16,18]. Type II saponites show a tendency of a small number of octahedral vacancies [15]. Examples of common formulae are:

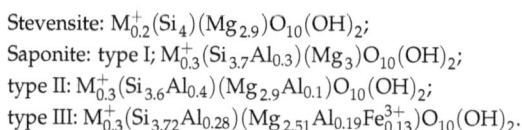

Stevensite: $M_{0.2}^{+}(Si_4)(Mg_{2.9})O_{10}(OH)_2$;
Saponite: type I; $M_{0.3}^{+}(Si_{3.7}Al_{0.3})(Mg_3)O_{10}(OH)_2$;
type II: $M_{0.3}^{+}(Si_{3.6}Al_{0.4})(Mg_{2.9}Al_{0.1})O_{10}(OH)_2$;
type III: $M_{0.3}^{+}(Si_{3.72}Al_{0.28})(Mg_{2.51}Al_{0.19}Fe_{0.13}^{3+})O_{10}(OH)_2$.

From the different layer structure of saponite and stevensite, the distance of charge to the layer surface is obviously different, which will determine hydration and sorption properties significantly [19,20].

The aim of the present study was the unambiguous characterization of a clay from the Madrid basin, which shows exceptional suitability as adsorbent material in biotechnology processes [21], and as adsorbent for mycotoxins [22]. In addition, pesticide removal from water has been demonstrated [23] for this clay. This clay can be also used to bind contaminants from the manufacture of paper [24]. Our study was initiated to better understand the structure—functionality relation of this material to potentially enhance its industrial use. The identification of smectite is important as saponite and stevensite vary in charge distribution influencing the absorption behavior of the material.

2. Material

Sample EX M 1694 (Clariant-internal distinct sample/clay quality, Clariant Produkte (Deutschland) GmbH, Frankfurt, Germany), a clay with a red-beige appearance from the Madrid basin, Spain) was studied to identify its mineralogical and chemical characteristics, especially the mineralogy of the interstratification.

3. Methods

Most methods to characterize 2:1 layer silicates were described in detail by Wolters et al. [25] and Steudel et al. [26]. Hence, only a brief description is provided here supplemented by detailed description of new methods. The methods were applied to the raw material as received and to the clay fraction (<2 μm; Na-exchanged), but the results here are focused primarily on the clay fraction.

3.1. Sample Preparation and Size Fractionation

The raw material (500 g) was divided by a rotating sample splitting device (rotary sample divider laborette 27, Fritsch, Idar-Oberstein, Germany) to obtain about 30 g of representative samples for mineralogical characterization. Chemical pre-treatments after Tributh and Lagaly [27] were applied to remove traces of carbonates, iron oxides, and organic matter [25]. Size fractionation was initiated to remove coarse crystallites of non-clay minerals and to enrich particles of <2 μm. For size separation, the material was first suspended in deionized water (800 mL) by mixing in an ultrasonic bath (30 min, Merck Eurolab, Darmstdt, Germany) and shaken overnight. The homogenous suspension was transferred into a 5 L beaker by passing a 63 μm sieve, which allowed separation of larger particles. The remaining suspension was diluted with deionised water to a solid content of about 1%. The <2 μm fraction was obtained by repeated gravitational sedimentation. The large volume was flocculated with NaCl (20× CEC, see below). Excess salt in the sediment was removed by dialysis (conductivity of surrounding deionized water <5 μS/cm). The dialysis tube (Nadir®, Carl Roth GmbH, Karlsruhe, Germany) consisted of cellulose hydrate with a width of 62.8 mm and a diameter of 40 mm. The chloride-free clay fraction was dried at 60 °C, gently manually ground (agate mortar) and stored in closed sample containers.

3.2. X-ray Diffraction (XRD) Analysis

The XRD patterns of the powdered bulk material and the clay fraction (<2 μm) were used for mineral identification and quantification. Mineral identification proceeded further by XRD patterns of oriented samples prepared from the clay fraction. Oriented samples were prepared by dispersing about 80 mg of the Na-exchanged <2 μm fraction in 2 mL deionized water. Samples were dried under atmospheric conditions at room temperature. After analyzing, the air dried samples were solvated for 48 h with ethylene glycol (EG) in a desiccator at 60 °C.

Measurements were performed with a Siemens D5000 (Bruker AXS GmbH, Karlsruhe, Germany) instrument equipped with a graphite diffracted-beam monochromator (CuKα, 40 kV, 40 mA, from 2°–45° 2θ, step width 0.02° 2θ, 3 s/step, divergence and antiscatter slit 0.6 mm, detector slit 1.0 mm). The mineral names were abbreviated according to Whitney and Evans [28].

3.3. Modeling of the One-Dimensional X-ray Pattern

"NEWMOD" was used to model one-dimensional X-ray pattern of different interstratifications to compare them with a measured XRD pattern of EG-solvated oriented specimen from our sample [29]. The EG-treated pattern was used to model the XRD data because the EG fixes the layer-to-layer space. In the air-dried state the basal spacing is sensitive to ambient humidity and the type of interlayer cations. Thus, it is difficult to determine the hydration state of the sample. Air-dried sample data are included here as a baseline to observe changes by EG solvation. A talc/smectite interstratification was selected for modeling. According to Moore and Reynolds [18], talc can be simulated by a trioctahedral mica model with zero values for K and Fe and by changing d(001) from 1.0 to 0.933 nm. A d(001) of 0.96 nm for turbostratic talc was applied during the modeling. For the swelling layers the following structure was selected: trioctahedral smectite-2Gly with a d(001) of 1.69 nm.

3.4. Infrared Spectroscopy—Attenuated Total Reflection Spectroscopy (ATR)

A Bruker IFS 55 EQUINOX spectrometer, equipped with a DTGS (deuterated triglycine sulphate) (Bruker Optik GmbH, Ettlingen, Germany) detector was employed to obtain IR-spectra. 64 scans in the 4000–400 cm^{-1} spectral range were recorded with a scanner velocity of 5 kHz and a resolution of 4 cm^{-1}. For the ATR measurements, a MIRacle single reflection diamond ATR cell (PIKE Technologies, Madison, WI, USA) was used. Sample preparation was simple: a small amount of powder was pressed on the diamond surface by a stainless steel-tipped anvil.

Band component analysis was undertaken using the Jandel Peakfit software package, (Version 4.12, Jandel Scientific, SeaSolve Software, Framingham, MA, USA), which enables the type of fitting function to be selected and allows specific parameters to be fixed or varied accordingly. The band fitting was done over a region from 1300 to 830 cm^{-1} using a Voigt function. A linear two-point background was chosen and fitting runs were repeated until reproducible results were obtained with a squared correlation parameter R^2 better than 0.998.

3.5. X-ray Fluorescence (XRF) Analysis

The chemical composition of the raw material and of the clay fraction (<2 µm; Na-exchanged) was determined by XRF using molten pellets with lithium tetraborate (mixing ratio 1:7). XRF analyses were performed on a Philips MagiXPRO spectrometer (PANalytical B.V., Almelo, The Netherlands, Company of Spectris plc., Egham, UK) equipped with a rhodium X-ray tube operated at 3.2 KW. The loss-on-ignition was determined prior to XRF by heating the samples at 1000 °C (2 h).

3.6. Simultaneous Thermal Analysis (STA)

The measurements were performed on a STA 449 C Jupiter (NETZSCH-Gerätebau GmbH, Selb, Germany) equipped with a thermogravimetric/differential scanning calorimetry (TG/DSC) sample holder. The STA is connected to a quadrupole mass spectrometer 403 C Aëolos (InProcess Instruments (IPI)/NETZSCH-Gerätebau GmbH, Selb, Germany) to detect the evolved gases from the sample during heating. All samples were allowed to equilibrate at a relative humidity (r.h.) of 53% in a desiccator above a saturated Mg(NO$_3$)$_2$ solution for at least 48 h. Conventional Pt/Rh crucibles (diameter 5 mm and height 5 mm) with a loosely-fitting perforated lid were filled with 80 mg of sample material. The measurements in the temperature range between 35 and 1100 °C with a heating rate of 10 K/min and an isothermal segment at 35 °C for 10 min were obtained under flowing synthetic air (SynA, 50 mL/min) mixed with nitrogen (20 mL/min) from the protective gas flow. The STA is connected to a pulse box (PulseTA, NETZSCH-Gerätebau GmbH, Selb, Germany) to inject a fixed volume of CO$_2$ (1 mL) in the flowing gas. Recording of the evolved CO$_2$ from sample and injected CO$_2$ allows the quantification of the evolved CO$_2$ e.g., from the oxidation of organic matter or from the decomposition of carbonates, [30]. Therefore, a 40 min isothermal segment is performed at the end of the measurement to inject CO$_2$ three times.

3.7. Cation Exchange Capacity (CEC) and Exchangeable Cations

To determine the CEC, the exchangeable cations were replaced by copper triethylenetetramine (Cu-trien) [31]. Prior to this exchange, the samples were also stored at 53% r.h. for at least 24 h. Fifty milligrams of the samples were dispersed in 5 mL 0.01 M Cu-complex solution. Ten milliliters of deionized water were added and the dispersions were shaken for 3 h. The dispersions were centrifuged once at a minimum of 3500 rpm for 10 min; afterwards, the clear blue supernatants (1 mL) were transferred into cuvettes (1.5 mL; semi-micro; disposable, Plastibrand, Brand GmbH & Co. KG, Wertheim, Germany). The depletion of the Cu-trien cations in the sample solutions was determined by photometry at 580 nm using a calibration series with different concentrations of Cu-trien. The CEC was related to the dry weight of the completely dehydrated material calculated with respect to water content determined by simultaneous thermal analysis (STA).

Exchangeable cations (Na$^+$, Ca^{2+}, K$^+$, Mg^{2+}) were determined with an optical emission spectrometer, where inductively-coupled plasma is used for excitation of the cations (ICP-OES; Jobin Yvon JY 38 S, HORIBA Europe GmbH, Oberursel, Germany). For the analysis, 5 mL of the supernatant solution from CEC determination were transferred with a pipette (Eppendorf GmbH,

Wesseling-Berzdorf, Germany) into a vial and diluted with deionized water in volume ratios of 1:1, 1:2, or 1:4 depending on the expected ion concentration from the CEC results. Ten to 20 µL of HNO_3 (65% supra pure) were added to decompose the Cu-trien complex and to stabilize the solutions.

3.8. Environmental Scanning Electron Microscope (ESEM)

Particle morphologies were examined by a Philips XL 30 FEG environmental scanning electron microscope (ESEM) (FEI Europe, Eindhoven, The Netherlands). Small amounts of sample powder were glued on aluminum SEM-holders using conductive tapes (Leit-C, Plano GmbH, Wetzlar, Germany). To improve the image quality, the samples were sputtered with a thin conductive layer (5 nm Au/Pd 80/20) and were investigated in low-vac mode at a chamber pressure of 1 Torr (130 Pa) using an acceleration voltage of 15 kV.

3.9. Specific Surface Area (A_S)

N_2 adsorption/desorption isotherms were recorded using either a Quantachrome Autosorb-1MP (Quantachrome Instruments, Boynton Beach, FL, USA) or an ASAP2010 (Micromeritics). To remove gaseous surface contaminants from the sample surfaces, the samples were held 24 h under vacuum at 95 °C prior to the gas sorption cycle. Specific surface area (A_S) was calculated by applying BET theory [32]. Six to 11 adsorption points in the range of p/p_0 from 0.01 to 0.2 were used for BET evaluation. The density functional theory (DFT) was applied to calculate micro- (<2 nm), meso- (2–50 nm) and macropore (>50 nm) volumes and areas (for definition of pore ranges see [33]) from the N_2 adsorption isotherm. For this purpose, the DFT kernel provided by the Quantachrome evaluation program was used. The DFT kernel is based on N_2 adsorbed at 77 K on silica surfaces with cylindrical pores and the non-local density functional theory (NLDFT) adsorption branch model [34]. The A_S was used to determine the range of square length (L) of the particles and layer (n) per stack [35–37].

4. Results and Discussion

4.1. Mineralogy

A broad diffraction peak below 10° (2θ) in the powder pattern of both the bulk sample (data not shown) and the clay fraction (Figure 1) revealed small coherent scattering domains. A 060 reflection at 0.152 nm revealed the existence of mainly trioctahedral clay minerals. No 060 reflection for dioctahedral clay minerals was observable). A small diffraction peak at the low angle side of the 0.319 nm reflection indicated traces of quartz in the bulk sample. Further impurities were calcite and gypsum, which could be proved by a peak in the mass spectrometer (MS) curve of evolved CO_2 ($m/z = 44$) and SO_2 ($m/z = 64$) during STA. The amount of calcite and gypsum was under the detection limit of XRD. The amount of the interstratification in the raw material and in the clay fraction was >97% and >99%, respectively. The powder pattern of the clay fraction showed typical hk reflections of the trioctahedral clay minerals at 0.455 nm and 0.257 nm (Figure 1). An additional reflection at 0.319 nm was observable, which was identified as the 005/003 reflection of the interstratification [9].

In the air-dried oriented pattern of the clay fraction (Figure 2) the first basal reflection is very broad and has a small shoulder at higher 2θ values, typical for interstratifications. Small particle sizes and turbostratic stacking produced line broadening in the X-ray pattern. After treatment with ethylene glycol, the broad diffraction peak splits in two diffraction peaks, due to the swelling of the smectite layers in the interstratification (Figure 2). Both the Na-exchanged and Ca-exchanged material showed after EG solvation the same expansion behavior (data not shown).

The FTIR spectra of the clay fraction of sample EX M 1694 were dominated by only a few features (Figure 3a). In detail, these features are a small band at 3675 cm^{-1}, which is at the same position as a talc-like Mg_3OH stretching vibration. The substitution of Al for Si, which transforms talc to Mg-rich smectite, leads to a remarkable broadening and to a slight displacement (5–10 cm^{-1}) of absorption bands of talc in the region below 1200 cm^{-1} (Figure 3a), which is in good agreement to Russel and

Fraser [38]. The very broad band between 3650 and 3000 cm^{-1} shows three maxima at 3621, 3420, and 3220 cm^{-1}. They can all be assigned to stretching vibrations of adsorbed water molecules on the clay surfaces [38]. The fingerprint region is dominated by a strong band at around 990 cm^{-1} (Figure 3a), having a number of shoulders on both sides. A spectrum deconvolution reveals four bands at 1092, 1057, 1015, and 973 cm^{-1} (Figure 3b), which can all be attributed to Si-O stretching modes, a common feature in clay mineral spectra [39].

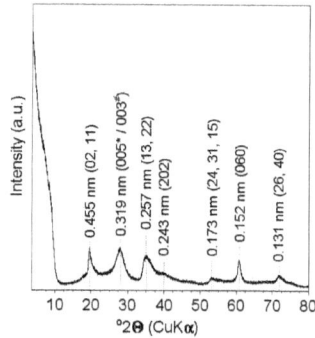

Figure 1. Powder pattern of the clay fraction <2 μm. In addition to the hkl reflections one 001 reflection of the interstratification contribution from (*) saponite and (#) turbostratic talc could be detected.

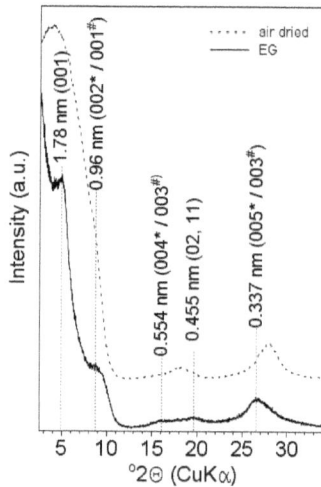

Figure 2. XRD patterns of the oriented clay fraction sample <2 μm. Non-rational basal reflections in the EG treated pattern with contributions from saponite (*) and turbostratic talc (#) are indicated.

The peak deconvolution reveals two more weak bands at 915 and 880 cm^{-1}. Both band positions are not reported for trioctahedral clay minerals, but are assigned to Al-OH-Al and Al-OH-Fe bending vibrations by Gates [40]. In addition, Cuadros et al. [10] attributed a band around 790 cm^{-1} in a talc-smectite interstratification to Fe-Mg-□-OH sites and a band around 800 cm^{-1} to Fe-Fe-□-OH sites [41].

The sharp band at 665 cm^{-1} is assigned to a Mg$_2$-R-OH bending mode with R = Mg for talc and R = Fe, Al for saponite. Its position is slightly lower than in talc (670 cm^{-1}) but higher than in saponite (655 cm^{-1}). This possibly occurs due to the interstratification of both types of layers and few substitutions in saponite layers.

The strong feature below 500 cm^{-1} seems to be a mixture of talc and saponite vibrations and is obviously not easy to assign. It has two maxima (440 and 421 cm^{-1}) and a shoulder at 464 cm^{-1}. This shoulder was assigned by Farmer [42] to a perpendicular Mg$_3$-OH bending vibration, while other authors attributed all bands to bending vibrations of the Si–O tetrahedra [43] or to vibrations of Si–O–Mg (464, 440 cm^{-1}) and Si–O units (421 cm^{-1}) [44].

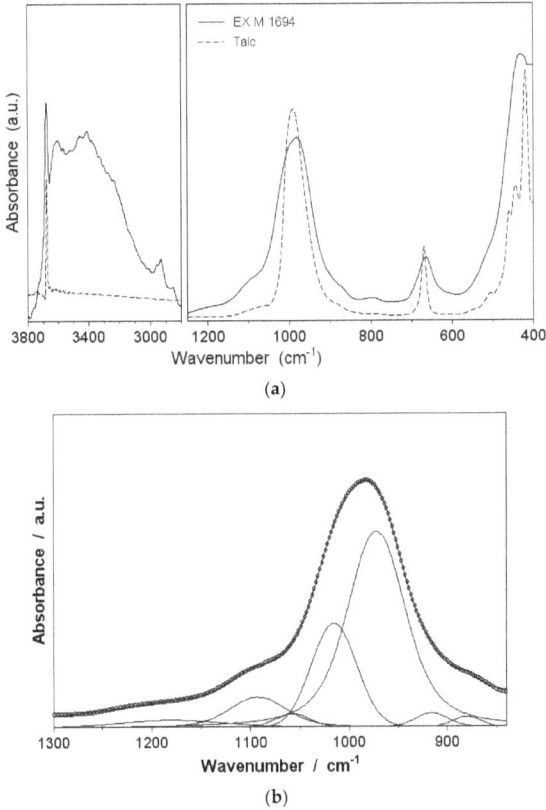

(a)

(b)

Figure 3. (a) Comparison of FTIR-ATR spectra of sample EX M 1694 and of talc; (b) Results of the spectrum deconvolution in the region between 830 and 1300 cm^{-1}. The band positions are at 880, 916, 973, 1015, 1057, 1092 cm^{-1}.

The differential scanning calorimetric (DSC) curves (Figure 4) show one endothermic peak in the region below 200 °C, which is associated with a maximum (130 °C) in the mass spectrometer (MS) curve of H$_2$O (m/z = 18), which reflects the release of adsorbed and interlayer water of a smectitic phase. Between 200 °C and 500 °C, a broad exothermic reaction occurs, which can be associated with the oxidation of some surface attached organic matter due to released H$_2$O and CO$_2$. Two further reactions were observed at higher temperatures (>500 °C) in the DSC curve. The first reaction was endothermic, but the peak maximum could not be determined in the DSC curve because the reaction is superposed by an exothermic reaction. The reactions are associated with gas emissions in the MS curve of H$_2$O (m/z = 18) and CO$_2$ (m/z = 44) and a mass loss of 4.1%. The H$_2$O release at 700 °C and above can be attributed to the dehydroxylation of trioctahedral clay minerals. Dehydroxylation (endothermal) occurs simultaneously with decomposition of clay minerals and recrystallization (exothermal) of high temperature phases. Thus, the maximum of the dehydroxylation peak (816 ± 1 °C) could only be determined by DTG and from the MS curve of evolved H_2O (m/z = 18). The small CO_2 release can be attributed to the decomposition of a carbonate, which also occurred simultaneously to the dehydroxylation of the clay mineral components. A small amount of carbonates (<1%) were determined by PulseTA.

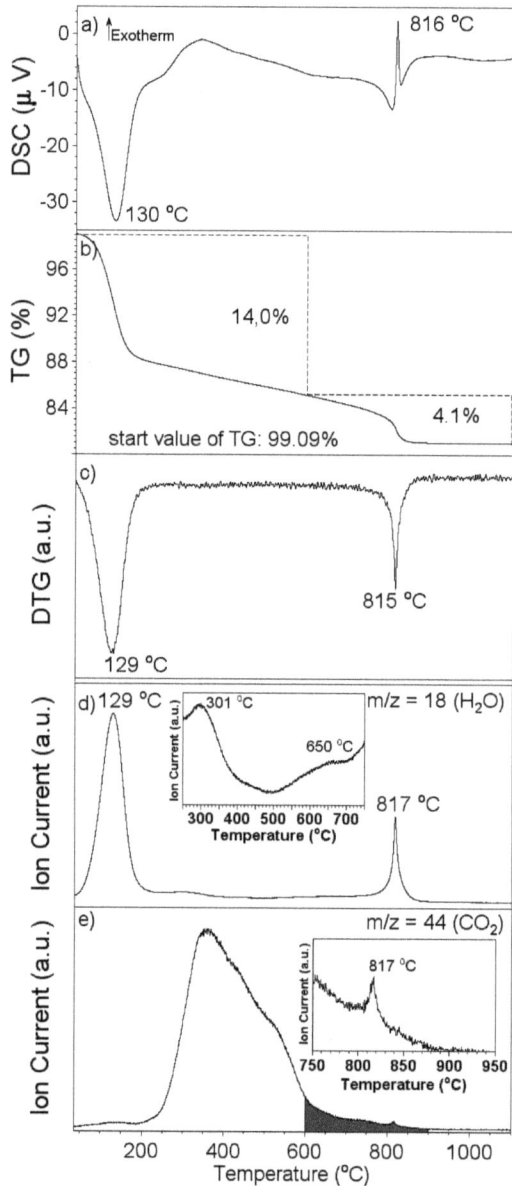

Figure 4. STA measurement of the clay fraction under SynA/N_2: (**a**) DSC curve; (**b**) TG curve; (**c**) DTG curve; (**d**) MS curve of H_2O; and (**e**) MS curve of CO_2.

Chemical analysis showed that the raw material and the clay fraction contain Si and Mg as main elements, which is consistent with the existence of Mg-rich clay mineral components, which are trioctahedral. The amount of Al was below 2% and the amount of Fe was below 1% (Table 1). The LOI of the raw material and for the clay fraction was 21.33 wt % and 18.08 wt %, respectively. The nickel content of the both materials averaged <10 ppm. The low Ni content is related to the composition of the parent rocks. The raw material has a Ca^{2+}/Mg^{2+} ratio of about 1:2 in the interlayer.

Table 1. Chemical composition of raw material and clay fraction (normalized to ignited conditions).

EX M 1694	SiO_2	Al_2O_3	MgO	Fe_2O_3	Na_2O	CaO	K_2O	Total
Raw material (wt %)	62.76	1.76	33.56	0.61	0.04	0.86	0.41	100
Clay fraction * (wt %)	63.82	1.84	32.20	0.55	1.36	0.04	0.18	100

* <2 μm Na-exchanged.

A comparison of ESEM images of the raw material and the clay fraction of EX M 1694 (Figure 5a,b) indicates, that the clay mineral particles formed small aggregates of about 4 μm and show floccules of micro aggregates. These aggregates show the typical fluffy or cloud-like smectite-morphology (Figure 5b).

Figure 5. ESEM images (**a**) distribution of micro aggregates in the raw material; and (**b**) typical fluffy smectite morphology of the particles in the clay fraction.

The layer charge had to be derived from the structural formula because layer charge measurements based on Lagaly [45] were not possible, owing to very broad peaks of low intensities in XRD patterns of n-alkylammonium exchanged samples. The specific surface area (A_S) of the raw material and the clay fraction is 278 m^2/g and 283 m^2/g, respectively. The A_S revealed three layers per stack ($n = 3$) and a very small lateral layer dimension between 60 and 80 nm.

4.2. Interstratification Studies

First, a turbostratic talc/smectite interstratification was modelled. The best result was reached using the following parameter ((Reichweite = 0, d(001) for turbostratic talc = 0.96 nm, smectite content = 0.7, LowN = 1; HighN = 5; δ = 2). Not all features in the XRD pattern could be reproduced with a turbostratic talc/smectite interstratification (Figure 6a). Therefore, a turbostratic talc pattern (Figure 6b) was modelled, too. The parameter LowN, HighN, and δ were the same as already chosen for the interstratification. The d(001) for the turbostratic talc was fixed to 0.96 nm. The two modelled patterns (Figure 6b) were mixed in the ratio: 75% turbostratic talc/smectite interstratification and 25% turbostratic talc. Figure 6c,d shows that all features could now be reproduced. The masked range (Figure 6a–d) of the pattern was excluded, because in this region the d(02,11) reflection of trioctahedral clay minerals is located [9]. The occurrence of this reflection indicated that the particles were not completely oriented parallel to the sample holder. The variation of the Fe content from 0 to 0.1 has no significant influence in the patterns, therefore the iron content in the smectite component was fixed to 0.05 p.f.u. according to the chemical composition (Table 1). HighN and δ were very low to reproduce the broad diffraction peaks. Higher values would produce very sharp diffraction peak profiles. The low values agreed with the results of BET measurements.

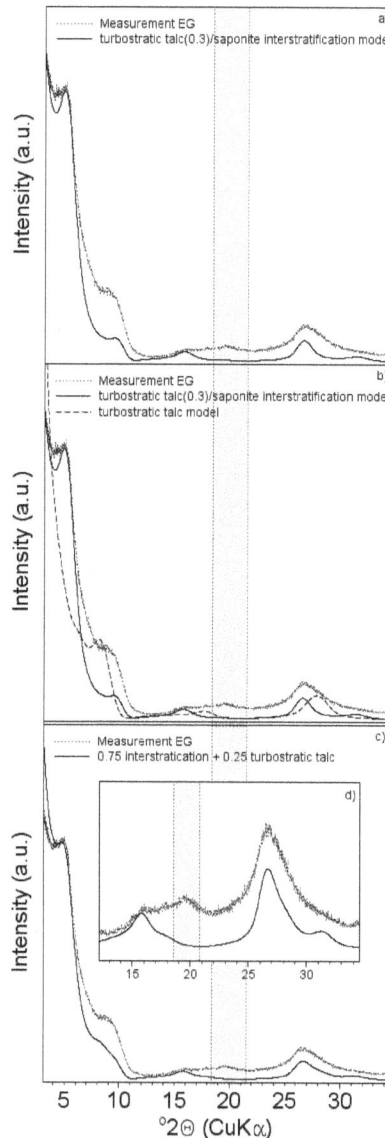

Figure 6. EG-treated XRD pattern of the oriented clay fraction sample in comparison to the modeled pattern by NEWMOD: (**a**) turbostratic talc (0.3)/saponite interstratification model; (**b**) turbostratic talc(0.3)/saponite interstratification model + turbostratic talc model; (**c**) mixture of interstratification (0.75) model + turbostratic talc model; and (**d**) enlarged view of (c).

According to the modeling results, the structural formula of the smectite in the interstratification was calculated [46] using the chemical analysis of the clay fraction (Table 1). Based on Cuadros et al. [10], we assumed that the chemical variability of turbostratic talc and talc is low compared to smectite. Thus, we applied the ideal talc formula $(Si_{4.0})(Mg_{3.0})O_{10}(OH)_2$ for both talc components. The structural formula for smectite (70%) was then calculated as: $M_{0.31}^+(Si_{3.77}Al_{0.23})(Mg_{2.84}Fe_{0.05}Al_{0.03})O_{10}(OH)_2$. M^+ includes the interlayer cations Na, Ca, and K. The determined structural formula for the smectite component is typical for a saponite type III rather than for stevensite, which was found by Martin de Vidales [3] in similar materials.

The H_2O release below 300 °C was higher than expected for a mixture of a Na-rich swelling clay mineral (average H_2O content at 53% r.h. at about 12.5%, unpublished data) and a fine-grained non-swelling clay mineral (maximum H_2O content at 53% r.h. of 5%). The high H_2O binding ability is caused by the extreme small material grains and large surface area.

To determine the mass loss resulting from dehydroxylation of the clay mineral components the overall mass loss between 600 °C and 1100 °C was corrected by the mass loss caused by thermal carbonate decomposition. Even 1% calcite would result in a mass loss of 0.44%, which would cause an overestimation of OH from overall mass loss in TG above 600 °C. Talc contains 4.6% water as hydroxyl groups and saponite 4.66%. The investigated sample consists of 75% turbostratic talc (0.3)/smectite interstratification and 25% turbostratic talc. Thus, the amount of turbostratic talc from the interstratification corresponds to 22.5% of total layers and the sum of turbostratic talc corresponds to 47.5%. The amount of saponite from the interstratification totaled 52.5% of all layers. The amount of OH from 47.5% talc is 2.19% and from 52.5% saponite is 2.45% (2.19% + 2.45% = 4.64%). The carbonate content in the clay sample fraction was 0.36%, which results in a mass loss of 0.16%. With the carbonate correction, the mass lost owing to dehydroxylation is 4.66% instead of 4.82% (uncorrected) and this is equivalent to the theoretical value. This result indicates no additional surface hydroxyl groups. The CEC of the raw material (37 cmol(+)/kg) is equal to the CEC of the clay fraction (40 cmol(+)/kg). The calculated layer charge of the pure smectite (0.31 p.f.u.; $M_{smectite}$ = 385.35 g/mol) revealed a CEC of 80 cmol(+)/kg. Thus, 52.5% saponite had a CEC of 42 cmol(+)/kg, which equates the measured CEC (40 cmol(+)/kg) of the clay fraction.

According to Faust & Murata [47], Veniale and Van der Marel [48], Grimshaw [49], Alietti and Mejsner [1], and Post [50], the endothermic decomposition of pure talc and saponite, as well as talc-saponite interstratifications involves crystallization of enstatite and cristobalite. Decomposition (endothermic) and recrystallization (exothermic) are out of the temperature range (up to 1100 °C) of conventional STA measurement systems [1,48–50]. In contrast, the investigated turbostratic talc-saponite interstratification showed an exothermic peak with a maximum at 815 °C superimposing with the endothermic dehydroxylation of the interstratification. The low temperature of dehydroxylation and decomposition for turbostratic talc-saponite interstratification is again caused by the small particle size.

4.3. Surface Properties of the Interstratification

The specific surface area of the raw material is 278 m^2/g and is only slightly lower than the specific surface area of the clay fraction at 283 m^2/g, and is comparable to engineered adsorption materials (Table 2, [51]).

Table 2. Surface and porosity characteristics of the talc-saponite interstratification in comparison to an acid-modified bentonite based on BET (p/p₀ = 0.01–0.2) and DFT calculations.

Parameters	EX M 1694 Bulk	EX M 1694 <2 μm	Ca-Saturated Bentonite Acid-Modified
A_s BET (m^2/g)	278	283	236
A_{MicroP} (m^2/g)	139	140	44
A_{MesoP} (m^2/g)	138	142	189
A_{MacroP} (m^2/g)	1	1	3

Engineered adsorption materials that are produced by acid treatment of smectitic clays to adjust surface area and porosity [52] generate hazardous waste. The studied material can be used as mined and, in addition, possesses a small amount of macropores, but large numbers of micropores (50%) and mesopores (50%). Therefore, it might not only be suitable for further (bio)-separation processes of low molecular substances without pretreatment, but also for adsorption of macromolecules, like proteins and enzymes.

5. Conclusions

Sample EX M 1694 from the Madrid basin (Spain) is similar to the pink clay from the same provenance studied by Martin de Vidales [3] and Cuevas et al. [4], but the interstratification consists of 30% turbostratic talc and 70% saponite type III rather than turbostratic talc and stevensite (due to tetrahedral and octahedral substitutions). In addition to the interstratification (75%), the sample contained 25% turbostratic talc. The fine grained natural material with small lateral layer dimension and low number of layers per stack possesses a large surface area comparable to engineered sorption materials. Therefore, it is highly suitable for application as mined without chemical pretreatment, which reduces the environmental burden. Furthermore, low particle size reduces thermal stability and dehydroxylation, and recrystallization takes place well below 1000 °C. The exceptional suitability of the studied material for absorbents in biotechnology processes, for mycotoxins and pesticides can be explained by its physical and chemical properties.

Acknowledgments: We are grateful to Katherina Rüping and Doreen Rapp for their help in the laboratory. The authors acknowledge Marita Heinle (Karlsruhe Institute of Technology, KIT) for ICP-OES analyses, Utz Kramar (Karlsruhe Institute of Technology, KIT) for XRF analyses and Peter Weidler (Karlsruhe Institute of Technology, KIT) for gas adsorption measurements. We are very grateful to Javier Cuadros, Steve Guggenheim, and two anonymous reviewer for discussions that improved the manuscript.

Author Contributions: Friedrich Ruf, Ulrich Sohling, Rainer Schuhmann, and Katja Emmerich conceived the project and designed the overall experimental strategy. Friedrich Ruf and Ulrich Sohling selected the sample. Annett Steudel performed the XRD, the STA and the CEC experiments and analyzed the data. Annett Steudel performed the modeling of the one-dimensional X-ray pattern. Frank Friedrich performed and evaluated the FTIR and ESEM experiments. All authors participated in writing the manuscript.

Conflicts of Interest: The authors declare no conflict of interest.

References

1. Alietti, A.; Mejsner, J. Structure of a talc/saponite mixed-layer mineral. *Clays Clay Miner.* **1980**, *28*, 388–390. [CrossRef]

2. Eberl, D.D.; Jones, G.; Khoury, H.N. Mixed-layer kerolite/stevensite from the Amargosa Desert, Nevada. *Clays Clay Miner.* **1982**, *57*, 115–133. [CrossRef]

3. De Vidales, J.L.M.; Pozo, M.; Alia, J.M.; Garcia-Navarro, F.; Rull, F. Kerolite-stevensite mixed-layers from the Madrid basin, Central Spain. *Clay Miner.* **1991**, *26*, 329–342. [CrossRef]

4. Cuevas, J.; Pelayo, M.; Rivas, P.; Leguey, S. Characterization of Mg-clays from the Neogene of the Madrid basin and their potential as backfilling and sealing material in high level radioactive waste disposal. *Appl. Clay Sci.* **1993**, *7*, 383–406. [CrossRef]

5. Pozo, M.; Casas, J. Origin of kerolite and associated Mg clays in palustrine-lacustrine environments. The Esquivias deposit (Neogene Madrid Basin, Spain). *Clay Miner.* **1999**, *34*, 395–418. [CrossRef]

6. De Santiago Buey, C.; Suarez Barrios, M.; Garcia Romero, E.; Dominiguez Diaz, M.C.; Doval Montoya, M. Electron microscopic study of the illite-smectite transformation in the bentonites from Cerro del Aquila (Toledo, Spain). *Clay Miner.* **1998**, *33*, 501–510. [CrossRef]

7. Brindley, G.W.; Bish, D.L.; Wan, H.-M. The nature of kerolite, its relation to talc and stevensite. *Mineral. Mag.* **1977**, *41*, 443–452. [CrossRef]

8. Bailey, S.W. Summary of recommendations of AIPEA nomenclature committee on clay minerals. *Am. Mineral.* **1980**, *65*, 1–7.

9. Brindley, G.W.; Brown, G. *Crystal Structures of Clay Minerals and Their X-ray Identification*; Monograph 5; Mineralogical Society: London, UK, 1980; p. 539.

10. Cuadros, J.; Vesselin, M.D.; Fiore, S. Crystal chemistry of the mixed-layer sequence talc-talc-smectite-smectite from submarine hydrothermal vents. *Am. Mineral.* **2008**, *93*, 1338–1348. [CrossRef]

11. Nickel, E.H.; Nichols, M.C. IMA/CNMNC List of Mineral Names. Available online: http://nrmima.nrm.se//MINERALlist.pdf (accessed on 22 May 2008).

12. Guggenheim, S. Introduction to Mg-rich clay minerals: Structure and composition. In *Magnesian Clays: Characterization, Origin and Applications*; Pozo, M., Galan, E., Eds.; AIPEA Educational Series: Bari, Italy, 2015; pp. 1–62.

13. Perdikatsis, B.; Burzlaff, H. Strukturverfeinerung am Talk $Mg_3[(OH)_2Si_4O_{10}]$. *Z. Kristallogr.* **1981**, *156*, 177–186. [CrossRef]

14. Guggenheim, S.; Adams, J.M.; Bain, D.C.; Bergaya, F.; Brigatti, M.F.; Drits, V.A.; Formoso, M.L.L.; Galán, E.; Kogure, T.; Stanjek, H. Summary of recommendations of nomenclature committees relevant to clay mineralogy: Report of the association Internationale pour l'etude des argiles (aipea) nomenclature committee for 2006. *Clays Clay Miner.* **2006**, *54*, 761–772. [CrossRef]

15. Newman, A.C.D. *Chemistry of Clays and Clay Minerals*; Monograph 6; Mineralogical Society: London, UK, 1987; p. 480.

16. Köster, H.M.; Schwertmann, U. Beschreibung einzelner Tonminerale. In *Tonminerale und Tone: Struktur, Eigenschaften, Anwendung und Einsatz in Industrie und Umwelt*; Jasmund, K., Lagaly, G., Eds.; Steinkopff Verlag: Darmstadt, Germany, 1993; pp. 33–88.

17. Mackenzie, R.C. Saponite from Allt Ribhein, Fiskavaig Bay, Skye. *Mineral. Mag.* **1957**, *31*, 672–680. [CrossRef]

18. Moore, D.M.; Reynolds, R.C., Jr. *X-ray Diffraction and the Identification and Analysis of Clay Minerals*; Oxford University Press: New York, NY, USA, 1997; p. 378.

19. Ferrage, E.; Sakharov, B.A.; Michot, L.J.; Lanson, B.; Delville, A.; Cuello, G.J. Water organization in Na-saponite: An experimental validation of numerical data. *Geochim. Cosmochim. Acta* **2010**, *74*, A289.

20. Ferrage, E.; Lanson, B.; Sakharov, B.A.; Geoffroy, N.; Jacquot, E.; Drits, V.A. Investigation of dioctahedral smectite hydration properties by modeling of X-ray diffraction profiles: Influence of layer charge and charge location. *Am. Mineral.* **2007**, *92*, 1731–1743. [CrossRef]

21. Temme, H.; Sohling, U.; Suck, K.; Ruf, F.; Niemeyer, B. Separation of aromatic alcohols and aromatic ketones by selective adsorption on kerolite-stevensite clay. *Colloids Surf. A* **2011**, *377*, 290–296. [CrossRef]

22. Sohling, U.; Haimerl, A. Use of Stevensite for Mycotoxin Adsorption. Patent WO 2006119967, 17 July 2012.

23. Ureña-Amate, M.D.; Socías-Viciana, M.; González-Pradas, E.; Saifi, M. Effects of ionic strength and temperature on adsorption of atrazine by a heat treated kerolite. *Chemosphere* **2005**, *59*, 69–74. [CrossRef] [PubMed]

24. Sohling, U.; Ruf, F. Stevensite and/or Kerolite Containing Adsorbents for Binding Interfering Substances during the Manufacture of Paper. Patent WO 200702941, 1 March 2007.

25. Wolters, F.; Lagaly, G.; Kahr, G.; Nüesch, R.; Emmerich, K. A comprehensive characterization of dioctahedral smectites. *Clays Clay Miner.* **2009**, *57*, 115–133. [CrossRef]

26. Steudel, A.; Batenburg, L.; Fischer, H.; Weidler, P.G.; Emmerich, K. Alteration of swellable clays by acid treatment. *Appl. Clay Sci.* **2009**, *44*, 105–115. [CrossRef]

27. Tributh, H.; Lagaly, G. Aufbereitung und Identifizierung von Boden- und Lagerstättentonen Teil I: Aufbereitung der Proben im Labor. *GIT Fachzeitschrift für das Laboratorium* **1986**, *30*, 524–529.

28. Whitney, D.L.; Evans, B.W. Abbreviations for names of rock-forming minerals. *Am. Mineral.* **2010**, *95*, 185–187. [CrossRef]

29. *NEWMOD for Windows*^{TM}*: The Calculation of One-Dimensional X-ray Diffraction Patterns of Mixed-Layered Clay Minerals*; Reynolds, R.C., Jr. and Reynolds, R.C., III.: Hanover, NH, USA, 1996; p. 25.

30. Emmerich, K. Thermal analysis in the characterization and processing of industrial minerals. In *Advances in the Characterization of Industrial Minerals*; EMU, Notes in Mineralogy: London, UK, 2011; pp. 129–170.

31. Meier, L.P.; Kahr, G. Determination of the cation exchange capacity (CEC) of clay minerals using the complexes of copper(II) ion with triethylenetetramine and tetraethylenepentamine. *Clays Clay Miner.* **1999**, *47*, 386–388. [CrossRef]

32. Brunauer, S.; Emmett, P.H.; Teller, E. Adsorption of gases in multimolecular layers. *J. Am. Chem. Soc.* **1938**, *60*, 309–319. [CrossRef]

33. Sing, K.S.W.; Everett, D.H.; Haul, R.A.W.; Moscou, L.; Pierotti, R.A.; Rouquerol, J.; Siemieniewska, T. Reporting physisorption data for gas/solid systems with special reference to the determination of surface area and porosity. *Pure Appl. Chem.* **1985**, *57*, 603–619. [CrossRef]
34. Lowell, S.; Shields, J.E.; Thomas, M.A.; Thommes, M. Characterization of porous solids and powders. In *Surface Area, Pore Size and Density*; Springer: Berlin, Germany, 2006; p. 347.
35. White, G.N.; Zelazny, L.W. Analysis and implications of the edge structure of dioctahedral phyllosilicates. *Clays Clay Miner.* **1988**, *36*, 141–146. [CrossRef]
36. Tournassat, C.; Neaman, A.; Villiéras, F.; Bosbach, D.; Charlet, L. Nanomorphology of montmorillonite particles: Estimation of the clay edge sorption site density by low-pressure gas adsorption and AFM observations. *Am. Miner.* **2003**, *88*, 1989–1995. [CrossRef]
37. Delavernhe, L.; Steudel, A.; Darbha, G.K.; Schäfer, T.; Schuhmann, R.; Wöll, C.; Geckeis, H.; Emmerich, K. Influence of mineralogical and morphological properties on the cation exchange behavior of dioctahedral smectites. *Colloids Surf. A* **2015**, *481*, 591–599. [CrossRef]
38. Russel, J.D.; Fraser, A.R. Infrared methods. In *Clay Mineralogy: Spectroscopic and Chemical Determinative Methods*; Wilson, M.J., Ed.; Chapman & Hall: London, UK, 1994; pp. 11–67.
39. Farmer, V.C. *The Infrared Spectra of Minerals*; Monograph 4; Mineralogical Society: London, UK, 1974; p. 331.
40. Gates, W.P. Infrared Spectroscopy and the Chemistry of Dioctahedral Smectites. In *The Application of Vibrational Spectroscopy to Clay Minerals and Layered Double Hydroxides*; Workshop Lectures; Kloprogge, J.T., Ed.; The Clay Mineral Society: Boulder, CO, USA, 2005; Volume 13, pp. 125–168.
41. Cuadros, J.; Altaner, S.P. Compositional and structural features of the octahedral sheet in mixed-layer illite/smectite from bentonites. *Eur. J. Miner.* **1988**, *10*, 111–124.
42. Farmer, V.C. The infra-red spectra of talc, saponite, and hectorite. *Mineral. Mag.* **1958**, *31*, 829–844. [CrossRef]
43. Kloprogge, J.T.; Frost, R.L. The effect of synthesis temperature on the FT-Raman and FT-IR spectra of saponites. *Vib. Spectrosc.* **2000**, *23*, 119–127. [CrossRef]
44. Van der Marel, H.W.; Beutelspacher, H. *Atlas of Infrared Spectroscopy of Clay Minerals and Their Admixtures*; Elsevier: Amsterdam, The Netherlands, 1976; p. 191.
45. Lagaly, G. Layer Charge Determination by Alkylammonium Ions. In *Layer Charge Characteristics of 2:1 Silicate Clay Minerals*; Mermut, A.R., Ed.; The Clay Minerals Society: Aurora, CO, USA, 1994; Volume 6, pp. 1–46.
46. Stevens, R.E. A system for calculating analyses of micas and related minerals to end members. *US Geol. Surv. Bull.* **1945**, *950*, 101–119.
47. Faust, G.T.; Murata, K.J. Stevensite, redefined as a member of the montmorillonite group. *Am. Mineral.* **1953**, *38*, 973–978.
48. Veniale, F.; Van der Marel, H.W. A regular Talc-Saponite mixed layer mineral from Ferriere, Nure Valley (Piacenza Province, Italy). *Contrib. Mineral. Petrol.* **1968**, *17*, 237–254. [CrossRef]
49. Grimshaw, R.W. *The Chemistry and Physics of Clays and Allied Ceramic Materials*, 4th ed.; Ernest Benn Limited: London, UK, 1971; p. 1024.
50. Post, J.L. Saponite from Near Ballarat, California. *Clays Clay Miner.* **1984**, *32*, 147–153. [CrossRef]
51. Sohling, U.; Ruf, F.; Schurz, K.; Emmerich, K.; Steudel, A.; Schuhmann, R.; Weidler, P.G.; Ralla, K.; Riechers, D.; Kasper, C.; et al. Natural mixture of silica and smectite as a new clayey material for industrial applications. *Clay Miner.* **2009**, *44*, 525–537. [CrossRef]
52. Komadel, P.; Madejová, J. Acid activation of clay minerals. In *Handbook of Clay Science*, 2nd ed.; Bergaya, F., Theng, B.K.G., Lagaly, G., Eds.; Elsevier: Amsterdam, The Netherlands, 2006; pp. 263–287.

minerals

MDPI

Article

Thermal Transformation of NH$_4$-Clinoptilolite to Mullite and Silica Polymorphs

Antonio Brundu, Guido Cerri * and Eleonora Sale

Department of Natural and Territorial Sciences, University of Sassari, Via Piandanna 4, 07100 Sassari, Italy; abrundu@uniss.it (A.B.); eleonorasale@hotmail.it (E.S.)
* Correspondence: gcerri@uniss.it; Tel.: +39-079-228621

Academic Editor: Annalisa Martucci
Received: 14 December 2016; Accepted: 13 January 2017; Published: 19 January 2017

Abstract: Clinoptilolite is a natural zeolite used for the abatement of ammonium in the treatment of urban wastewater. By considering that mullite was obtained through thermal treatment of NH$_4$-exchanged synthetic zeolites, this work aimed to evaluate if this phase can be obtained from NH$_4$-clinoptilolite. A material containing about 90 wt % of clinoptilolite, prepared using a Sardinian zeolite-rich rock, was NH$_4$-exchanged and subjected to treatments up to 1200 °C. After dehydration, de-ammoniation, and dehydroxylation processes, the clinoptilolite structure collapsed at 600 °C. An association of mullite, silica polymorphs, and glass, whitish in color, was obtained for treatments between 1000 and 1200 °C. The higher degree of crystallinity was reached after a 32 h heating at 1100 °C: mullite 22 wt %, cristobalite 59 wt %, tridymite 10 wt %, glass 9 wt %. It is possible to speed up the kinetics of the transformation by increasing the temperature to 1200 °C, obtaining the same amount of mullite in 2 h, but increasing the residual amorphous fraction (16 wt %). These results indicate that NH$_4$-clinoptilolite could represent a raw material of potential interest in the ceramic field, in particular in the production of acid refractory, opening scenarios for a possible reuse of clinoptilolite-based exchangers employed in ammonium decontamination.

Keywords: zeolite; clinoptilolite; mullite; ammonium; ammonia; cristobalite; ceramic; refractory; thermal treatment; waste

---

1. Introduction

Clinoptilolite is the most abundant among natural zeolites and high-grade deposits are distributed worldwide [1]. Not only has it cation exchange capacity, but it also exhibits high selectivity toward NH$_4^+$, known since the sixties of the past century, when the first studies, addressed to exploit this feature in the treatment of municipal wastewater, were accomplished [2]. However, the use of clinoptilolite in ammonium decontamination still remains a matter of interest [3–5]. In the United States, nearly 80% of the zeolites sold in the domestic market is related to uses exploiting ammonia/ammonium adsorption, such as animal feed, odor control, water purification, pet litter, and wastewater treatment [6]; noteworthy, clinoptilolite represents more than 85% of US production [7]. Clinoptilolite has been employed in the treatment of urban wastewater for the removal of NH$_4^+$ [1,8], also taking advantage of its low cost, although only a limited number of plants, three in the USA and fourteen in Australia, have operated [1]. It should be noted that the regeneration of the exhausted zeolite, as well as the recovery of ammonia, are feasible processes, but often not cost-effective [9]. This general rule has resulted in being pushed to find uses for the spent exchangers, for example clinoptilolite containing ammonium ions can be used as fertilizer [8], and should encourage further studies aimed to evaluate new alternatives.

Heating determines transformations in the structure of zeolites, and some general rules governing the correlation between the composition, original framework, and the thermal stability of these minerals have been established [10]. It has been demonstrated that some NH_4-exchanged synthetic zeolites can be transformed into an association of mullite and amorphous silica by thermal treatments [11–13]. Mullite has achieved outstanding importance as a material for both traditional and advanced ceramics because of its favorable thermal and mechanical properties [14]. The use of natural zeolites in ceramic production has been evaluated by different research groups, highlighting the advantages, generally a lowering of sintering temperatures and limits, mainly linked to the dark color often observed in the fired products [15–19]. Recent papers show that crystalline materials can be obtained by thermally induced transformation of Cs- and Pb-exchanged clinoptilolite [20–22], whereas only an amorphous phase has been achieved from the thermal treatment of a NH_4-clinoptilolite [23].

The continuous increase of quantity of inorganic waste has stimulated, as a challenge, different studies designed to transform waste into resources for the ceramic industry [24–26]. With this perspective, a material containing NH_4-clinoptilolite, derived from a wastewater treatment, might be evaluated as a potential raw material for the ceramic industry. On the basis of the above mentioned considerations, the present research has been addressed to evaluate the possibility of obtaining a ceramic matrix by heating an NH_4-exchanged clinoptilolite.

2. Experimental Section

2.1. Starting Material and Beneficiation Process

The present research was performed by using a clinoptilolite-bearing epiclastite (sample labeled as "LacBen") [27], collected in the valley of the Tirso River (Northern Sardinia, Italy). Literature data report zeolite contents that span from 66 to 70 wt % for this material [22,28].

The rock was subjected to the beneficiation process described in previous papers [20,22,29,30], aimed to increase the zeolite content. Briefly, the material was submitted to autogenous comminution and dry sieving. Then, the fraction below 100 μm was subjected to ultrasound attack and wet separation in deionized water. The obtained powder was dried at 70 °C in a ventilated drying oven, then conditioned for 24 h at 22 ± 3 °C and 53% ± 5% of Relative Humidity (hereafter, RH), monitored with an Ebro Data Logger EBI20-TH1 (Ebro, Ingolstadt, Germany), using a desiccator containing a saturated solution of $Ca(NO_3)_2$. The material so obtained was labeled ES-AR.

2.2. Preparation of NH_4-Clinoptilolite

To obtain a NH_4-clinoptilolite, ES-AR was previously Na-exchanged, a procedure that allows an improvement of the cation exchange capacity [31]. The enriched powder was contacted with a 1 M NaCl solution (Merck ACS salt; purity > 99.5%) performing a sequence of ten exchange cycles of 2 h each, executed in a batch at 65 °C under continuous stirring, with a solid/liquid ratio of 30 g/L. The last two exchange cycles were performed using a VWR Prolabo salt (purity 99.9%). The Na-exchanged material was rinsed with deionized water until complete removal of chloride solution (test performed on elutes with $AgNO_3$). The powder was dried at 35 °C overnight, then rehydrated for 24 h at 22 °C and 53% ± 2% of RH. The Na-clinoptilolite was conducted in NH_4-form using a 0.5 M NH_4Cl solution (Sigma Aldrich salt; purity 99.5%), by carrying out five exchange cycles using the same conditions of Na-preconditioning. Once rinsed, the material was dried and rehydrated as described above. The NH_4-exchanged material was labeled ES-NH.

2.3. Chemical Analysis

The chemical analysis of ES-NH was performed at the Activation Laboratories Ltd (Ancaster, ON, Canada). Major elements were determined after lithium metaborate/tetraborate fusion of the sample through Inductive Coupled Mass Atomic Emission Spectrometry (ICP-AES), performed with a Varian Vista 735 ICP (Varian, Inc., Palo Alto, CA, USA). NH_4 content was calculated on the basis of the Total N determined through the Total Kjeldahl Nitrogen (TKN) method. The Loss of Ignition (LoI) of the material was determined, in duplicate, by calcination of the sample for 2 h at 1000 °C. The H_2O

content in the NH_4-clinoptilolite was calculated as the difference between the LoI and the $(NH_4)_2O$ content [32].

2.4. Thermal Treatments

Aliquots of 250 mg of ES-NH were submitted to thermal treatments of 2 h at 200, 300, 400, 500, 600, 700, 800, 900, 1000, 1100, and 1200 °C, performed in a muffle furnace (Vittadini mod. FS. 3, Vittadini, Milano, Italy) using platinum crucibles. Further experiments were performed at 1000 and 1100 °C, at each temperature heating aliquots of 250 mg of sample for 4, 8, 16, and 32 h.

2.5. X-ray Diffraction (XRD)

ES-AR, ES-NH, and all heated samples were investigated employing a Bruker D2-Phaser (Bruker, Karlsruhe, Germany) with the following conditions: 30 kV, 10 mA, CuKα radiation, LynxEye detector with an angular opening of 5°, 2θ range 6°–70°, step size 0.020°, time per step 2 s, spinner 15 rpm. Before the measurements, all the samples were micronized using a Retsch MM400 mill (ZrO_2 cups and balls). ES-AR, ES-NH heated for 32 h at 1000 and 1100 °C, and ES-NH heated for 2 h at 1200 °C were also analyzed by adding to the specimens 20 wt % of corundum as internal standard. All measurements were performed using a low-background silicon crystal specimen holder (Bruker), except for the ES-AR, placed in a steel sample holder (Bruker). The XRD patterns were evaluated using the software EVA 4.1.1 (2015; Bruker DIFFRACplus Package) coupled with the database PDF-2 (ICDD). Quantitative analyses were performed with the Rietveld method using the software Bruker Topas 4.5.

2.6. Thermal Analyses

Thermogravimetric, Derivative Thermogravimetric and Differential Thermal Analyses (hereafter, TG, DTG, and DTA) of ES-NH were carried out using a TA Instrument Q600 (TA Instruments, New Castle, DE, USA) simultaneous thermal analyzer. Amounts of about 15 mg of sample were heated up to 1300 °C, both under air (five analyses) and nitrogen flow (one analysis; N_2 purity 99.999%, Sapio), in an alumina crucible at the following operating conditions: 10 °C/min; gas flow 100 mL/min. The software TA-Universal Analysis was used to evaluate the results.

2.7. Scanning Electron Microscope (SEM) Observations

Morphological observations were carried out on the samples heated for 32 h at 1100 °C and for 2 h at 1200 °C. The materials, placed on aluminum stubs, were gold coated by sputtering and observed using a ZEISS DSM 962 Scanning Electron Microscope (Zeiss, Oberkochen, Germany).

3. Results

The mineralogical composition of ES-AR is reported in Table 1 (the Rietveld refinement is provided in Figure S1, Supplementary Materials). The beneficiation process enabled a powder to be obtained with a clinoptilolite content of about 89 wt %, along with residual amounts of feldspars, glass, opal-CT, biotite, and quartz. This result confirms that the beneficiation process here adopted is effective and replicable [20,22,29,30].

Table 1. Mineralogical composition of sample ES-AR (values in wt %; e.s.d. = estimated standard deviation; R-weighted pattern (Rwp) = 6.96%).

ES-AR	Clinoptilolite	Feldspars	Quartz	Opal-CT	Biotite	Amorphous	Sum
content	89.3	4.0	0.7	2.1	1.0	2.9	100.0
e.s.d.	±4.0	±1.0	±0.2	±0.5	±0.2	±1.2	-

The chemical composition of ES-NH is reported in Table 2.

Table 2. Chemical composition of sample ES-NH (values in wt %).

SiO$_2$	Al$_2$O$_3$	Fe$_2$O$_3$	MnO	MgO	CaO	Na$_2$O	K$_2$O	TiO$_2$	P$_2$O$_5$	(NH$_4$)$_2$O	H$_2$O	Sum
67.85	12.81	0.78	0.01	0.39	0.30	0.14	0.46	0.23	0.05	5.76	10.92	99.70

The high ammonium content (2.21 meq/g), along with the low sodium, potassium, calcium, and magnesium contents, indicate that a near end-member of NH$_4$-clinoptilolite was obtained.

The XRD patterns of ES-NH and of all the samples heated for 2 h from 200 to 1200 °C are reported in Figure 1. The diffractograms show that the structure of clinoptilolite is well recognizable also after the treatment at 500 °C, although a slight shifting of the peaks toward higher 2θ angles, and a reduction of their intensities, occurred. The heating at 600 °C determined the amorphization of the material (Figure 1a), and no further change was recorded up to 1000 °C, when the nucleation of cristobalite began (Figure 1b). The XRD pattern of ES-NH treated at 1100 °C shows, beside cristobalite, also traces of tridymite and mullite.

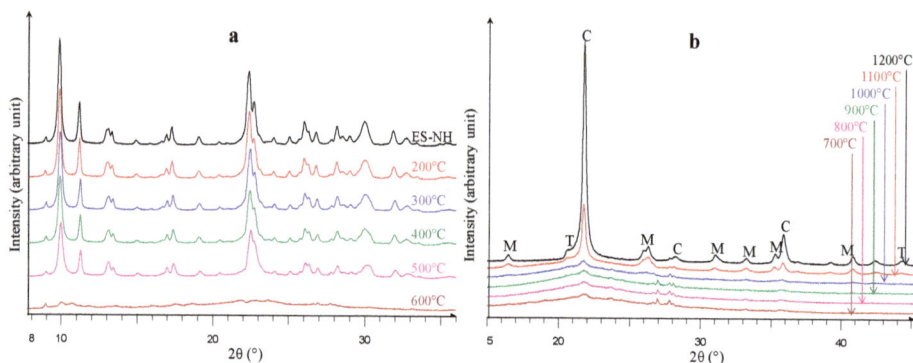

Figure 1. (**a**) X-ray Diffraction (XRD) patterns of ES-NH unheated and treated for 2 h from 200 to 600 °C and; (**b**) from 700 to 1200 °C. C = Cristobalite; M = Mullite; T = Tridymite.

At 1200 °C, a matrix almost entirely crystalline, basically composed of cristobalite (54.3 wt %), tridymite (8.8 wt %) and mullite (21.0 wt %), was obtained (Figure 1b and Table 3).

Table 3. Mineralogical compositions of ES-NH heated at the temperatures and for the time indicated (values in wt %).

Temperature	Time (h)	Cristobalite	Tridymite	Mullite	Amorphous	Sum	Rwp
1000 °C	32	5.1	3.5	5.2	86.2	100.0	3.91
e.s.d.	-	±1.0	±0.7	±1.0	±6.0	-	-
1100 °C	32	59.3	10.1	21.8	8.8	100.0	7.97
e.s.d.	-	±2.8	±1.6	±1.8	±2.0	-	-
1200 °C	2	54.3	8.8	21.0	15.9	100.0	8.18
e.s.d.	-	±2.5	±1.4	±1.8	±2.5	-	-

The XRD patterns of the ES-NH heated up to 32 h at 1000 and 1100 °C are reported in Figure 2. In both cases the crystallization increases with the time, but very slowly at 1000 °C, indeed after 32 h the amorphous phase is largely dominant (about 86 wt %—Table 3). Conversely, the residual glassy fraction is just 8.8 wt % in the sample heated at 1100 °C for 32 h, that is mainly composed of cristobalite (about 59 wt %) and mullite (almost 22 wt %), as reported in Table 3 (the Rietveld refinement is provided in Figure S2, Supplementary Materials). A thermal treatment at 1200 °C allows 21 wt % of mullite to be obtained in just 2 h, but in this case the amorphous fraction reaches 16 wt % (Table 3).

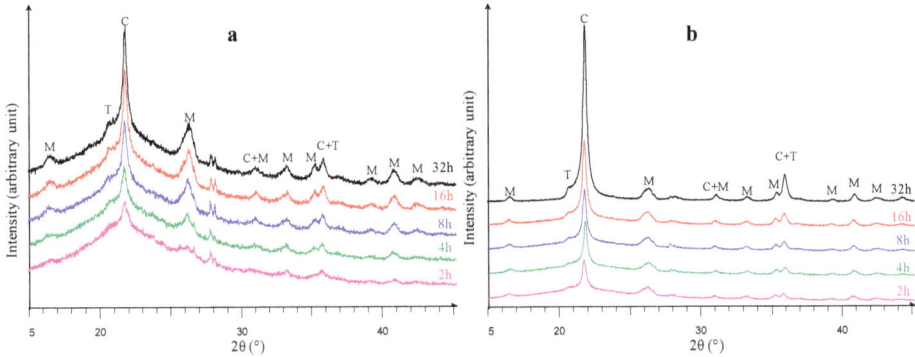

Figure 2. (**a**) XRD patterns of ES-NH treated up to 32 h at 1000; and (**b**) at 1100 °C. C = Cristobalite; M = Mullite; T = Tridymite.

The results of the thermal analyses of ES-NH are reported in Figure 3. The weight loss occurred in four steps, well-marked in the DTG curves. Except for the slight offset of the last weight loss, the TG and DTG curves follow the same paths, regardless of the gas used during the analysis; no mass loss was recorded above 800 °C. Besides the endothermic peaks associated to the weight losses, the DTA curves show three exothermic reactions, at 550, 1041, and 1149 °C when the sample was analyzed under an air flow, and at 577, 1014, and 1148 °C in a nitrogen atmosphere. Finally, both DTA curves exhibit a broad endotherm at about 1200 °C.

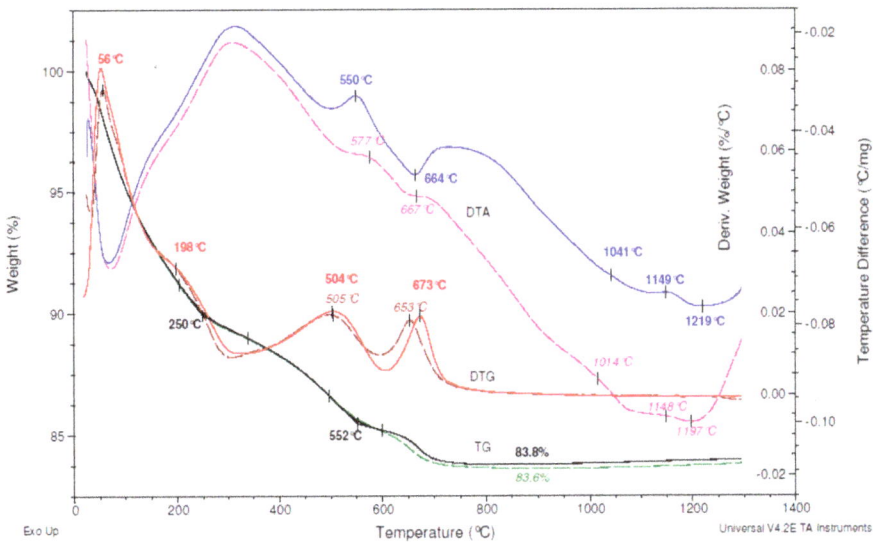

Figure 3. Thermogravimetric, Derivative Thermogravimetric, and Differential Thermal Analyses (TG-DTG-DTA) curves of ES-NH analyzed under flow of air (solid lines) or nitrogen (dashed lines).

SEM observations performed on ES-NH heated at 1200 °C evidenced rounded grains, often coalescent (Figure 4a); at higher magnification, the presence of acicular shapes (about 1 μm in length) can be inferred (Figure 4b). These morphologies were not noticed in the sample treated at 1100 °C for 32 h.

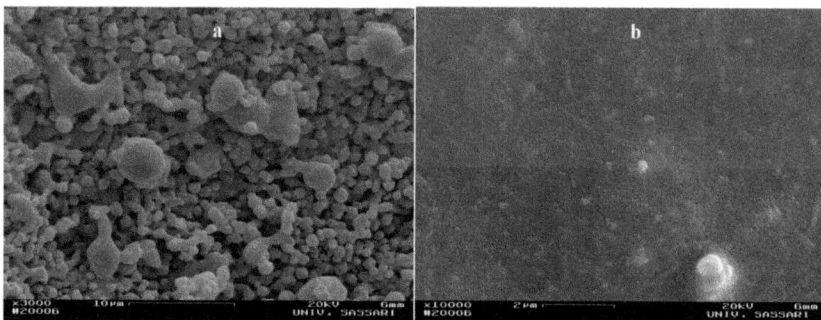

Figure 4. Scanning electron microscopy (SEM) images of ES-NH heated at 1200 °C.

4. Discussion

In discussing the phenomena occurring during the heating of ES-NH it is appropriate to start with the thermal analysis, and compare the results with those reported by Tomazovic et al. [32] in their detailed study on the properties of a Serbian clinoptilolite conducted in NH_4-form. In Figure 3, the peaks marked on the DTG curve at 56 and 198 °C are related to the loss of water. At about 250 °C the material starts to evolve NH_3, a process that overlaps to the residual dehydration still in progress. From the kinetic point of view, the weight loss associated with ammonia release reaches the maximum at about 500 °C (see DTG), and this process ends at about 550 °C (see offset on the TG curve), substantially in agreement with the data in the literature [32]. A heating time of 2 h at 500 °C should have been sufficient to evolve all the ammonia contained in ES-NH, hence the corresponding XRD pattern in Figure 1a can be attributed to an H-form of clinoptilolite [33]. The slight shifting of several peaks toward higher 2θ angles, observable by comparing the XRD patterns of ES-NH heated up to 500 °C (Figure 1a), is compatible with a progressive reduction of the cell volume of clinoptilolite, determined by the dehydration and de-ammoniation processes [32].

The exothermic peak at 550 °C along the DTA path in Figure 3 (see analysis under air flow), is related to the combustion of the ammonia released from the zeolite. In air, this phenomenon should take place at 651 °C, but clinoptilolite (like other zeolites) can catalyze the auto-ignition of ammonia, triggering this process already at 530–570 °C [32,34]. When the analysis was performed under a flow of nitrogen, because of the reduced availability of oxygen inside the furnace, the exothermic peak of ammonia combustion showed weaker intensity, and a shift of almost 30 °C toward higher temperatures (Figure 3).

The DTG curves of ES-NH show a sharp peak at 650–675 °C (Figure 3), linked to the dehydroxylation of the H-clinoptilolite [33]. This weight loss is also marked by an evident endothermic reaction along the DTA paths (at 664–667 °C, Figure 3), and the five analyses performed under air flow showed always the same peaks in the same positions. These DTG/DTA peaks are not present in the thermal analyses of NH_4-clinoptilolite previously reported by other authors [32–34]; this may be due to a combination of the following factors: (i) the content of clinoptilolite in the material; (ii) the content of NH_4^+ in the zeolite; (iii) the performance of the thermal analyzer used. The reaction of dehydroxylation accompanies the collapse of the zeolite framework. In fact after a heating time of 2 h at 600 °C, ES-NH becomes amorphous (Figure 1a), whereas the breakdown of the NH_4-clinoptilolite prepared by Tomazevic et al. [23] begins at 600 °C, a difference that can be explained by the higher Si/Al ratio of the Serbian clinoptilolite (5.02) [32] with respect to that of the Sardinian zeolite contained in ES-NH (4.71) [28].

Above 1000 °C, the DTA curves show a first weak exothermic (more difficult to recognize in the analysis performed under air flow) between 1014 and 1041 °C, and a second, clearer, at about 1150 °C (Figure 3). Such peaks are attributable to the nucleation of cristobalite and mullite, respectively, in fact, in the XRD patterns of Figure 1b, the main reflection of the silica polymorph becomes distinguishable after a treatment of 2 h at 1000 °C, whereas the nucleation of mullite required a heating at 1100 °C. The broad endothermic peak along the DTA curves in Figure 3, with a minimum at about 1200 °C, should correspond to the formation of a liquid phase.

The nucleation of cristobalite from thermally treated zeolites is not a novelty [22,23,35] but, so far, it has never been reported for NH_4-clinoptilolite. An analogous consideration applies to mullite; in fact, literature reports the thermal transformation of NH_4-exchanged synthetic zeolites to mullite [12,13], but this phase has never been obtained from natural zeolites. It should be noted that Tomazovic et al. [23] heated a NH_4-clinoptilolite for 2 h at 1100 °C without obtaining mullite or cristobalite.

XRD results summarized in Figure 2 and Table 3 indicate that the kinetic aspects of the transformation from NH_4-clinoptilolite to mullite and silica polymorphs are relevant. At 1000 °C the reaction proceeds very slowly, remaining largely incomplete after 32 h, whereas with the same time at 1100 °C enabled the best result in terms of crystallinity to be obtained: only 8.8 wt % of amorphous fraction. It is possible to speed up the transformation by increasing the temperature to 1200 °C, but this results in an increase of the residual glassy fraction (15.9 wt %).

With respect to SEM observations performed on ES-NH heated for 2 h at 1200 °C, the shape of the grains and their coalescence could be due to incipient melting (Figure 4a), an hypothesis consistent with the results of thermal analyses, whereas the (rare) needle-like morphologies could correspond to mullite crystals (Figure 4b). In spite of an almost identical mineralogical composition, the sample heated for 32 h at 1100 °C does not show these morphologies, probably because the phase transformations took place in the solid state. On the other hand, this is the case for a material containing 63 wt % of mullite obtained from zeolite A [13], that shows the morphology of the precursor even if the zeolite structure has been destroyed. The phases nucleated from ES-NH cannot inherit the habitus of clinoptilolite because it is destroyed during the enrichment process [20].

Results show that during the heating, but before the nucleation of the high-temperature phases, NH_4-clinoptilolite undergoes the phenomena summarized by Jacobs et al. [33], here schematized for a clinoptilolite with Si/Al ratio = 5:

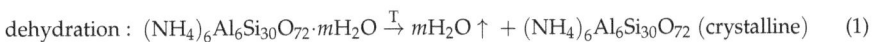

$$\text{dehydration}: \ (NH_4)_6Al_6Si_{30}O_{72}\cdot mH_2O \xrightarrow{T} mH_2O \uparrow \ + (NH_4)_6Al_6Si_{30}O_{72} \ (\text{crystalline}) \qquad (1)$$

$$\text{de-ammoniation}: \ (NH_4)_6Al_6Si_{30}O_{72} \xrightarrow{T} 6NH_3 \uparrow + H_6Al_6Si_{30}O_{72} \ (\text{crystalline}) \qquad (2)$$

$$\text{dehydroxylation}: \ H_6Al_6Si_{30}O_{72} \xrightarrow{T} 3H_2O \uparrow + Al_6Si_{30}O_{69} \ (\text{amorphous}) \qquad (3)$$

Such phenomena, accompanied by the mass losses detected through thermogravimetric analysis (Figure 3), determine a progressive transformation in the chemical composition of ES-NH (Table 4), illustrated by using the M_2O-Al_2O_3-SiO_2 ternary diagram in Figure 5.

Table 4. Chemical composition of ES-NH after dehydration (D-1), de-ammonion (D-2) and dehydroxilation (D-3), with mullite (Mul) composition calculated from the formula $Al_6Si_2O_{13}$ (wt %).

Sample	SiO_2	Al_2O_3	Fe_2O_3	MnO	MgO	CaO	Na_2O	K_2O	TiO_2	P_2O_5	M_2O	Sum
D-1	76.43	14.43	0.88	0.01	0.44	0.34	0.16	0.52	0.26	0.06	6.48 [a]	100.00
D-2	79.81	15.07	0.92	0.01	0.46	0.35	0.17	0.54	0.27	0.06	2.34 [b]	100.00
D-3	81.73	15.43	0.94	0.01	0.47	0.36	0.17	0.56	0.27	0.06	-	100.00
Mul	28.20	71.80	-	-	-	-	-	-	-	-	-	100.00

[a] M corresponds to ammonium; [b] M corresponds to hydrogen.

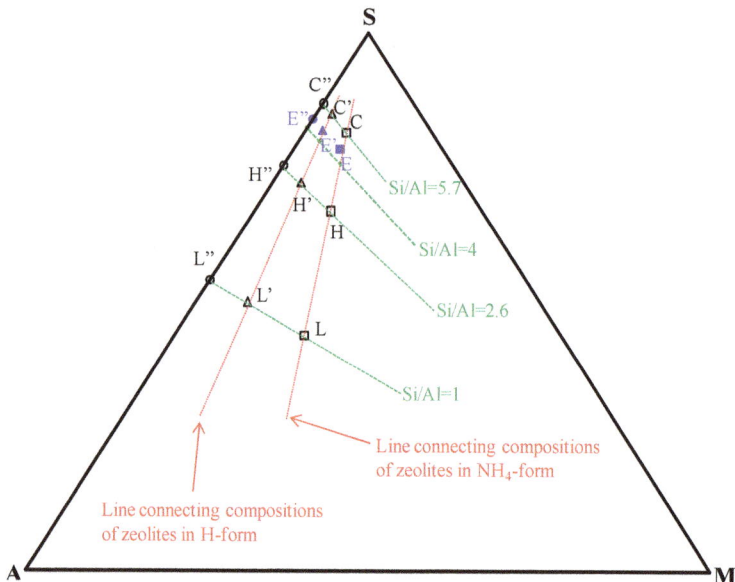

Figure 5. Ternary plot for M_2O-Al_2O_3-SiO_2 system (values in wt %). S = silica; A = alumina; M = oxide of ammonium (for square marks) or hydrogen (for triangle marks). C = dehydrated NH_4-clinoptilolite (Si/Al = 5.7); E = dehydrated ES-NH; H = dehydrated NH_4-heulandite (Si/Al = 2.6); L = dehydrated NH_4-LTA zeolite. C′, E′, H′, and L′: composition of the same materials after de-ammoniation. C″, E″, H″, and L″: composition of the same materials after dehydroxylation.

The components of the system are the oxides of: silicon at the vertex S, aluminum at the vertex A, and, at vertex M, alternatively ammonium (for square marks as E, that corresponds to ES-NH dehydrated) or hydrogen (for triangle marks as E′, that corresponds to de-ammoniated ES-NH). Once dehydroxylated, the composition of ES-NH is referable to the binary system Al_2O_3-SiO_2 (circle marked E″ in Figure 5). To evaluate the approximation of this plot, it can be noted that the sum of oxides not belonging to the system corresponds to 2.66, 2.78, and 2.84 wt % for ES-NH dehydrated, de-ammoniated and dehydroxylated, respectively (Table 4). These approximations are in line with those common in evaluating raw materials for the ceramic industry [36] (p. 193). Being substantially constituted by a clinoptilolite with a Si/Al ratio = 4.71, ES-NH has a Si/Al ratio = 4.53, hence the points relative to this material (E, E′, E″ in Figure 5) fall above the line corresponding to Si/Al = 4, used to distinguish heulandite (Si/Al ratio < 4) from clinoptilolite [37]. Moreover, the point E is located close to the line connecting the theoretical compositions of zeolites in NH_4-form, thus confirming that a near-end-member of NH_4-clinoptilolite was obtained. The point E′ in Figure 5 ideally corresponds to ES-NH after ammonia evolution, i.e., to a material containing clinoptilolite in H-form [33]. After dehydroxylation, the chemistry of ES-NH reaches the composition SiO_2 = 84.1 wt %; Al_2O_3 = 15.9 wt % (E″ in Figure 5). In the binary system silica-alumina reported by Manning [36] (p. 191), this composition falls in the field of "fireclays", raw materials that are used to produce acid refractories constituted by an association of mullite and cristobalite. In the Al_2O_3-SiO_2 system the eutectic temperature is 1590 °C, but the impurities in ES-NH (2.84 wt %; Table 4) determine the formation of a liquid phase at about 1200 °C, as shown by the thermal analyses (Figure 3).

An association constituted by mullite and amorphous silica was obtained by thermal treatment of two synthetic zeolites (X and LTA), previously NH_4-exchanged [12,13]. Among zeolites, LTA has the lowest Si/Al ratio (Si/Al = 1). In the diagram of Figure 5 the compositional variation of a NH_4-LTA zeolite after dehydration, de-ammoniation and dehydroxylation is reported. The same compositional pathways were drawn for the two ammonium forms at the limits of the heulandite-clinoptilolite series, having Si/Al ratios of 2.6 and 5.7, respectively [38]. Once amorphized, all the zeolites reported in the diagram of Figure 5 reach compositions that, on the silica-alumina side, correspond to the field of "fireclays" [36] (p. 191).

By comparing the results of the present research with the data reported by Kosanovic et al. [12], it can be deduced that the Si/Al ratio affects the temperature of mullite nucleation and the kinetics of the transformation. In fact, NH_4-exchanged LTA (Si/Al = 1) and X (Si/Al = 1.22) synthetic zeolites complete the reaction after 3 h at 1000 °C [12], whereas the NH_4-clinoptilolite contained in ES-NH (Si/Al = 4.71) requires a treatment of 32 h at 1100 °C. This hypothesis is reinforced by the thermal behavior of a NH_4-clinoptilolite with a Si/Al ratio of 5.01 which, heated for 2 h at 1100 °C, was not transformed to crystalline phases [23], and the same result was obtained after a 3 h treatment at 1000 °C of a NH_4-mordenite with a Si/Al ratio of 4.35 [12].

The general reaction that leads to the formation of mullite and silica polymorphs from NH_4-clinoptilolite can be summarized as:

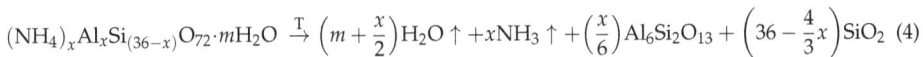

$$(NH_4)_x Al_x Si_{(36-x)} O_{72} \cdot m H_2 O \xrightarrow{T} \left(m + \frac{x}{2}\right) H_2 O \uparrow + x NH_3 \uparrow + \left(\frac{x}{6}\right) Al_6 Si_2 O_{13} + \left(36 - \frac{4}{3}x\right) SiO_2 \quad (4)$$

with x that spans from 5.4 to 7.2 (and from 5.4 to 10 by considering the whole heulandite–clinoptilolite series) [38].

From the data in Table 4, with a simple proportion it is possible to calculate that the percentage of $Al_6 Si_2 O_{13}$ theoretically obtainable from ES-NH corresponds to 21.5 wt %. This value is consistent with the highest content determined through XRD analysis (21.8 wt %, Table 3). On the other hand, mullite shows various Si/Al ratios referring to the solid solution $Al_{4+2x} Si_{2-2x} O_{10}$, with x ranging between about 0.2 and 0.9, and its structure is able to incorporate a number of transition metal cations and other foreign atoms [14]. Hence, the percentage of mullite (impure) could be slightly higher than the amount calculated on the basis of the theoretical formula $Al_6 Si_2 O_{13}$. Moreover, crystalline phases, having formulae that deviate from ideal compositions, were already obtained by thermal treatment of synthetic zeolites [39] or clinoptilolite [21,22].

Previous studies demonstrated that some zeolite-rich rocks can be used to partially substitute the feldspathic fluxes commonly employed in porcelain stoneware production, though a dark color represents a frequent drawback [15,16,19]. In general, clinoptilolite-rich rocks have evidenced better performances with respect to those containing other natural zeolites, like phillipsite and chabazite [15]. The results of this research are interesting, because they indicate that NH_4-clinoptilolite might represent a raw material of potential interest in the ceramic field, in particular in the production of acid refractory. Above all, it should be considered that clinoptilolite is a cheap commodity employable in NH_4^+ decontamination [1,8], and such a process would provide a material that can be transformed to an association of mullite and silica polymorphs, which means finding a way to turn waste into a resource. Finally, the synthesized matrix shows a whitish color (Figure S3, Supplementary Materials), a valuable feature in the ceramic field.

Further investigations should clarify the lowest clinoptilolite content and the minimum ammonium content necessary to obtain the transformation from NH_4-clinoptilolite to mullite and silica polymorphs.

Supplementary Materials: The following are available online at www.mdpi.com/2075-163X/7/1/11/s1. Figure S1: Experimental (blue) and calculated (red) X-ray powder diffraction pattern for ES-AR. Figure S2: Experimental (blue) and calculated (red) X-ray powder diffraction pattern for ES-NH heated for 32 h at 1100 °C. Figure S3: Sample of ES-NH recovered after the thermal analysis (air flow).

Acknowledgments: This work was partially financed by the Fund "Analisi Mineralogiche Cerri", provided by Guido Cerri. Authors are grateful to two anonymous reviewers whose observations have improved the manuscript.

Author Contributions: Antonio Brundu and Guido Cerri conceived and designed the experiments; Antonio Brundu and Eleonora Sale performed all the experiments, except for the thermal analyses, executed by Guido Cerri; Antonio Brundu and Guido Cerri analyzed the data; Guido Cerri provided reagents/materials/analysis tools; Antonio Brundu and Guido Cerri wrote the manuscript.

Conflicts of Interest: The authors declare no conflict of interest.

References

1. Eyde, T.H.; Holmes, D.A. Zeolites. In *Industrial Minerals and Rocks: Commodities, Markets, and Uses*, 7th ed.; Kogel, J.E., Trivedi, N.C., Barker, J.M., Krukowski, S.T., Eds.; Society for Mining, Metallurgy, and Exploration, Inc.: Littleton, CO, USA, 2006; pp. 1039–1064.
2. Pabalan, R.; Bertetti, F.P. Cation-Exchange Properties of Natural Zeolites. In *Natural Zeolites: Occurrence, Properties, Applications—Reviews in Mineralogy and Geochemistry*, 1st ed.; Bish, D.L., Ming, D.W., Eds.; Mineralogical Society of America: Washington, DC, USA, 2001; Volume 45, pp. 453–518.
3. Jorgensen, T.C.; Weatherley, L.R. Ammonia removal from wastewater by ion exchange in the presence of organic contaminants. *Water Res.* **2003**, 37, 1723–1738. [CrossRef]
4. Sprynskyy, M.; Lebedynets, M.; Terzyk, A.P.; Kowalczyk, J.; Namieśnik, J.; Buszewski, B. Ammonium sorption from aqueous solutions by the natural zeolite Transcarpathian clinoptilolite studied under dynamic conditions. *J. Colloid Interface Sci.* **2005**, 284, 408–415. [CrossRef] [PubMed]
5. Mazloomi, F.; Jalali, M. Ammonium removal from aqueous solutions by natural Iranian zeolite in the presence of organic acids, cations and anions. *J. Environ. Chem. Eng.* **2016**, 4, 240–249. [CrossRef]
6. Virta, R.L.; Flanagan, D.M. Zeolites. In *Minerals Yearbook 2014*; U.S. Geological Survey: Reston, VA, USA, 2015; Volume 1, pp. 83.1–83.3.
7. Flanagan, D.M. Zeolites. In *Mineral Commodity Summaries 2016*; U.S. Geological Survey: Reston, VA, USA, 2016; pp. 190–191.
8. Armbruster, T. Clinoptilolite-heulandite: Applications and basic research. *Stud. Surf. Sci. Catal.* **2001**, 135, 13–27.
9. Liberti, L.; Boghetich, G.; Lopez, A.; Petruzzelli, D. Application of microporous materials for the recovery of nutrients from wastewaters. In *Natural Microporous Materials in Environmental Technology—Nato Science Series*, 1st ed.; Misaelides, P., Macášek, F., Pinnavaia, T.J., Colella, C., Eds.; Springer Science & Business Media, B.V.: Dordrecht, The Netherlands, 1999; Volume 362, pp. 253–270.
10. Cruciani, G. Zeolites upon heating: Factors governing their thermal stability and structural changes. *J. Phys. Chem. Solids* **2006**, 67, 1973–1994. [CrossRef]
11. Matsumoto, T.; Goto, Y.; Urabe, K. Formation process of mullite from NH_4^+-exchanged Zeolite A. *J. Ceram. Soc. Jpn.* **1995**, 103, 93–95. [CrossRef]
12. Kosanović, C.; Subotić, B.; Smit, I. Thermally induced phase transformations in cation-exchanged zeolites 4A, 13X and synthetic mordenite and their amorphous derivatives obtained by mechanochemical treatment. *Thermochim. Acta* **1998**, 317, 25–37. [CrossRef]
13. Kosanović, C.; Subotić, B. Preparation of mullite micro-vessels by a combined treatment of zeolite A. *Microporous Mesoporous Mater.* **2003**, 66, 311–319. [CrossRef]
14. Schneider, H.; Schreuer, J.; Hildmann, B. Structure and properties of mullite—A review. *J. Eur. Ceram. Soc.* **2008**, 28, 329–344. [CrossRef]
15. De Gennaro, R.; Cappelletti, P.; Cerri, G.; de' Gennaro, M.; Dondi, M.; Guarini, G.; Langella, A.; Naimo, D. Influence of zeolites on the sintering and technological properties of porcelain stoneware tiles. *J. Eur. Ceram. Soc.* **2003**, 23, 2237–2245. [CrossRef]
16. De Gennaro, R.; Dondi, M.; Cappelletti, P.; Cerri, G.; de' Gennaro, M.; Guarini, G.; Langella, A.; Parlato, L.; Zanelli, C. Zeolite-feldspar epiclastic rocks as flux in ceramic tile manufacturing. *Microporous Mesoporous Mater.* **2007**, 105, 273–278. [CrossRef]
17. Demirkiran, A.Ş.; Artir, R.; Avci, E. Effect of natural zeolite addition on sintering kinetics of porcelain bodies. *J. Mater. Process. Tech.* **2008**, 203, 465–470. [CrossRef]
18. Ergul, S.; Sappa, G.; Magaldi, D.; Pisciella, P.; Pelino, M. Microstructural and phase transformations during sintering of a phillipsite rich zeolitic tuff. *Ceram. Int.* **2011**, 37, 1843–1850. [CrossRef]

19. Sokolář, R.; Šveda, M. The use of zeolite as fluxing agent for whitewares. *Procedia Eng.* **2016**, *151*, 229–235. [CrossRef]

20. Brundu, A.; Cerri, G. Thermal transformation of Cs-clinoptilolite to $CsAlSi_5O_{12}$. *Microporous Mesoporous Mater.* **2015**, *208*, 44–49. [CrossRef]

21. Gatta, G.D.; Brundu, A.; Cappelletti, P.; Cerri, G.; de' Gennaro, B.; Farina, M.; Fumagalli, P.; Guaschino, L.; Lotti, P.; Mercurio, M. New insights on pressure, temperature, and chemical stability of $CsAlSi_5O_{12}$, a potential host for nuclear waste. *Phys. Chem. Miner.* **2016**, *43*, 639–647. [CrossRef]

22. Brundu, A.; Cerri, G. Release of lead from Pb-clinoptilolite: managing the fate of an exhausted exchanger. *Int. J. Environ. Sci. Technol.* **2016**. [CrossRef]

23. Tomazović, B.; Ćeranić, T.; Sijarić, G. The properties of the NH_4-clinoptilolite. Part 2. *Zeolites* **1996**, *16*, 309–312. [CrossRef]

24. Andreola, F.; Barbieri, L.; Karamanova, E.; Lancellotti, I.; Pelino, M. Recycling of CRT panel glass as fluxing agent in the porcelain stoneware tile production. *Ceram. Int.* **2008**, *34*, 1289–1295. [CrossRef]

25. Dondi, M.; Guarini, G.; Raimondo, M.; Zanelli, C. Recycling PC and TV waste glass in clay bricks and roof tiles. *Waste Manag.* **2009**, *29*, 1945–1951. [CrossRef] [PubMed]

26. Andreola, F.; Barbieri, L.; Lancellotti, I.; Leonelli, C.; Manfredini, T. Recycling of industrial wastes in ceramic manufacturing: State of art and glass case studies. *Ceram. Int.* **2016**, *42*, 13333–13338. [CrossRef]

27. Cerri, G.; Cappelletti, P.; Langella, A.; de' Gennaro, M. Zeolitization of Oligo-Miocene volcaniclastic rocks from Logudoro (northern Sardinia, Italy). *Contrib. Mineral. Petrol.* **2001**, *140*, 404–421. [CrossRef]

28. Cerri, G.; de' Gennaro, M.; Bonferoni, M.C.; Caramella, C. Zeolites in biomedical application: Zn-exchanged clinoptilolite-rich rock as active carrier for antibiotics in anti-acne topical therapy. *Appl. Clay Sci.* **2004**, *27*, 141–150. [CrossRef]

29. Brundu, A.; Cerri, G.; Colella, A.; de Gennaro, M. Effects of thermal treatments on Pb-clinoptilolite. *Rend. Online Soc. Geol. Italy* **2008**, *3*, 138–139.

30. Cerri, G.; Farina, M.; Brundu, A.; Daković, A.; Giunchedi, P.; Gavini, E.; Rassu, G. Natural zeolites for pharmaceutical formulations: Preparation and evaluation of a clinoptilolite-based material. *Microporous Mesoporous Mater.* **2016**, *223*, 58–67. [CrossRef]

31. Kesraoul-Oukl, S.; Cheeseman, C.; Perry, R. Effects of conditioning and treatment of chabazite and clinoptilolite prior to lead and cadmium removal. *Environ. Sci. Technol.* **1993**, *27*, 1108–1116. [CrossRef]

32. Tomazović, B.; Ćeranić, T.; Sijarić, G. The properties of the NH_4-clinoptilolite. Part 1. *Zeolites* **1996**, *16*, 301–308. [CrossRef]

33. Jacobs, P.A.; Uytterhoeven, J.B.; Beyer, H.K.; Kiss, A. Preparation and Properties of Hydrogen Form of Stilbite, Heulandite and Clinoptilolite Zeolites. *J. Chem. Soc. Faraday Trans. 1* **1979**, *75*, 883–891. [CrossRef]

34. Langella, A.; Pansini, M.; Cerri, G.; Cappelletti, P.; de Gennaro, M. Thermal behavior of natural and cation-exchanged clinoptilolite from Sardinia (Italy). *Clays Clay Min.* **2003**, *51*, 625–633. [CrossRef]

35. Stoch, L.; Waclawska, I. Phase Transformations in Amorphous Solids. *High Temp. Mater. Proc.* **1994**, *13*, 181–201. [CrossRef]

36. Manning, D.A.C. *Introduction to Industrial Minerals*, 1st ed.; Chapman & Hall: London, UK, 1995; p. 276.

37. Coombs, D.S.; Alberti, A.; Armbruster, T.; Artioli, G.; Colella, C.; Galli, E.; Grice, J.D.; Liebau, F.; Mandarino, J.A.; Minato, H.; et al. Recommended nomenclature for zeolite minerals: Report of the Subcommittee on Zeolites of the International Mineralogical Association, Commission on New Minerals and Mineral Names. *Can. Miner.* **1997**, *35*, 1571–1606.

38. Bish, D.L.; Boak, J.M. Clinoptilolite-Heulandite Nomenclature. In *Natural Zeolites: Occurrence, Properties, Applications—Reviews in Mineralogy and Geochemistry*, 1st ed.; Bish, D.L., Ming, D.W., Eds.; Mineralogical Society of America: Washington, DC, USA, 2001; Volume 45, pp. 207–216.

39. Radosavljevic-Mihajlovic, A.S.; Kremenovic, A.S.; Dosen, A.M.; Andrejic, J.Z.; Dondur, V.T. Thermally induced phase transformation of Pb-exchanged LTA and FAU-framework zeolite to feldspar phases. *Microporous Mesoporous Mater.* **2015**, *201*, 210–218. [CrossRef]

minerals

MDPI

Article

The Influence of the Framework and Extraframework Content on the High Pressure Behavior of the GIS Type Zeolites: The Case of Amicite

Rossella Arletti [1,2], Carlotta Giacobbe [3], Simona Quartieri [4] and Giovanna Vezzalini [5,*]

[1] Dipartimento di Scienze della Terra, Università di Torino, Via Valperga Caluso 35, 10125 Torino, Italy; rossella.arletti@unito.it

[2] CrisDI Interdipartemental Center, University of Torino, Via Giuria 7, 10125 Torino, Italy

[3] ESRF-European Synchrotron Radiation Facility, CS 40220-38043 Grenoble Cedex 9, Grenoble, France; giacobbe@esrf.fr

[4] Dipartimento di Scienze Matematiche e Informatiche, Scienze Fisiche e Scienze della Terra, Università di Messina, Viale Ferdinando Stagno d'Alcontres 31, 98166 Messina S. Agata, Italy; squartieri@unime.it

[5] Dipartimento di Scienze Chimiche e Geologiche, Università di Modena e Reggio Emilia, Via G. Campi 103, 41125 Modena, Italy

* Correspondence: mariagiovanna.vezzalini@unimore.it; Tel.: +39-059-205-8471

Academic Editor: Annalisa Martucci
Received: 30 November 2016; Accepted: 29 January 2017; Published: 5 February 2017

Abstract: This paper reports a study, performed by in-situ synchrotron X-ray Powder Diffraction, of the high pressure behavior of the natural zeolite amicite [$K_4Na_4(Al_8Si_8O_{32})\cdot10H_2O$], the GIS-type phase with ordered (Si, Al) and (Na, K) distribution. The experiments were carried out up to 8.13(5) GPa in methanol:ethanol:water = 16:3:1 (m.e.w.) and 8.68(5) GPa in silicone oil (s.o.). The crystal structure refinements of the patterns collected in m.e.w. were performed up to 4.71(5) GPa, while for the patterns collected in s.o. only the unit cell parameters were determined as a function of pressure. The observed framework deformation mechanism—similar to that reported for the other studied phases with GIS topology—is essentially driven by the distortion of the "double crankshaft" chains and the consequent changed shape of the 8-ring channels. The pressure-induced over-hydration observed in the experiment performed in aqueous medium occurs without unit cell volume expansion, and is substantially reversible. A comparison is made with the high pressure behavior of the other GIS-type phases, and the strong influence on compressibility of the chemical composition of both framework and extraframework species is discussed.

Keywords: zeolite; amicite; high pressure; compressibility; in-situ synchrotron XRPD; pressure-induced hydration (PIH); structure refinement

1. Introduction

In the last 15–20 years, studies on the behavior of both natural and synthetic microporous materials under high pressure (HP) have multiplied noticeably, providing not only important information on their elastic behavior and stability, but also opening new perspectives for technological applications. For instance, among the physical properties of microporous materials investigated under compression, worthy of mention are: the so called P-induced amorphization processes (PIA) (e.g., [1–6]), the effect of pressure on the ionic conductivity (e.g., [7,8]), the P-induced over-hydration (PIH) (e.g., [9–14]) and the penetration of gas, like Ar, Xe, and CO_2 [15–17]. High pressure experiments on porous materials have recently led to the synthesis of linear carbon based polymers in pure silica zeolites. Linear polymers like polyacetylene (PA), polyethylene (PE), and polycarbonyl (pCO) have been obtained from compression resulting in nanocomposite organic/inorganic materials, which are good candidates for developing highly directional semiconductors and high energy materials [18].

The HP behavior of zeolites when compressed in non-penetrating fluids has recently been reviewed by Gatta and Lee (2014) [19] and summarized in the following way: (i) microporosity does not necessarily imply high compressibility, in fact the range of compressibility is wide, with bulk modulus K_0 ranging from ~15 to ~70 GPa; (ii) the flexibility observed in zeolites is based mainly on tetrahedra tilting; (iii) the deformation mechanisms are dictated by the framework topology; (v) the extraframework content (cations and water molecules) governs the compressibility level in isotypic structures.

Zeolites with GIS topology [20] and GIS-like materials have been studied under both high temperature and high pressure, revealing widely variable degrees and mechanisms of deformation as a function of the non-ambient experimental conditions and the chemical composition of both the framework and extraframework. The study of gismondine dehydration [21] showed that this framework is particularly flexible.

The HP behavior of a natural gismondine was studied using both "non-penetrating" (i.e., silicone oil, s.o.) [22] and "penetrating" (methanol:ethanol:water = 16:3:1, m.e.w.) [11] pressure-transmitting media (PTM). In the latter case, a PIH effect was observed at a very low P, inducing full occupation of originally partially occupied water sites. On the whole, both experiments revealed an unexpected low compressibility of gismondine, notwithstanding the high flexibility showed by this framework during dehydration and the similar framework deformation mechanisms [21].

Lee et al. [23] and Jang et al. [24] studied the HP behavior of two synthetic phases with GIS topology, both compressed in penetrating media: a K-gallosilicate (K-GaSi-GIS) and a K-aluminogermanate (K-AlGe-GIS), respectively. These studies highlighted a very different response to hydrostatic pressure in materials sharing the same GIS topology, but with considerably different framework and extraframework compositions.

Two microporous mixed octahedral-pentahedral-tetrahedral (OPT; [25]) framework silicates, structurally related to the GIS topology, were studied under HP [26]: cavansite and pentagonite, the orthorhombic dimorphs of $Ca(VO)(Si_4O_{10})\cdot4H_2O$. When compressed in m.e.w., these two phases exhibit rather different behaviors: pentagonite undergoes PIH, thanks to the crucial role of the seven-fold coordinated Ca, suitable for accepting an additional H_2O molecule. In contrast, in cavansite the eight-fold coordinated Ca cations do not allow further water penetration and thus PIH is not observed. The higher compressibility in s.o. of cavansite compared to gismondine is attributed to the presence of VO_5 pyramids connecting the tetrahedral layers of the vanadosilicate.

This paper presents a study, performed by in-situ synchrotron X-ray Powder Diffraction (XRPD), of the HP stability and behavior of the natural zeolite amicite $[K_4Na_4(Al_8Si_8O_{32})\cdot10H_2O]$, the GIS phase with ordered (Si, Al) and (Na, K) distribution. The investigation aimed in particular to understand: (i) the relationships between compressibility and framework/extraframework content; (ii) the influence of different PTM (penetrating 16:3.1 m.e.w. and non-penetrating s.o., respectively) on the compressibility and HP deformation mechanisms of this zeolite.

2. Amicite Structure

Amicite [ideal formula $K_4Na_4(Al_8Si_8O_{32})\cdot10H_2O$] is a rare natural zeolite, classified as the ordered K, Na member of the gismondine group [27]. The sample used for this study is from the type locality (Höwenegg in Hegau, southern West Germany)—where amicite was discovered associated with merlinoite in a basaltic rock—and is the same studied by Alberti and Vezzalini [28] (chemical formula: $K_{3.75}Na_{3.61}Ca_{0.05}[Al_{7.86}Si_{8.24}O_{32}]\cdot9.67H_2O$). Its GIS framework topology is shared by the other natural zeolites gismondine, garronite, gobbinsite, and by several other synthetic phases. Amicite structure [28] was determined in the monoclinic *I*2 s.g. The cell parameters are $a = 10.226(1)$, $b = 10.422(1)$, $c = 9.884(1)$ Å, $\beta = 88°$ 19(1). The framework can be described as intersecting ribbons of 4-membered rings of tetrahedra (defined as double-crankshaft chains) running in the *a* and *c* directions

(Figures 1 and 2), laterally linked to form two sets of channels delimited by 8-membered rings running parallel to [100] and [001]. The ordered distribution of Si and Al in the tetrahedra, and of Na and K in the channels, induces a lowering in symmetry from the topological $I4_1/amd$ space group to the real one $I2$. Na and K are distributed in two different and fully occupied sites, with the water molecules in four sites, three of which are fully occupied (W1, W2, W3). Na is coordinated to three framework oxygen atoms and to all the water molecules, while K is coordinated to four framework oxygen atoms and the three fully occupied water sites.

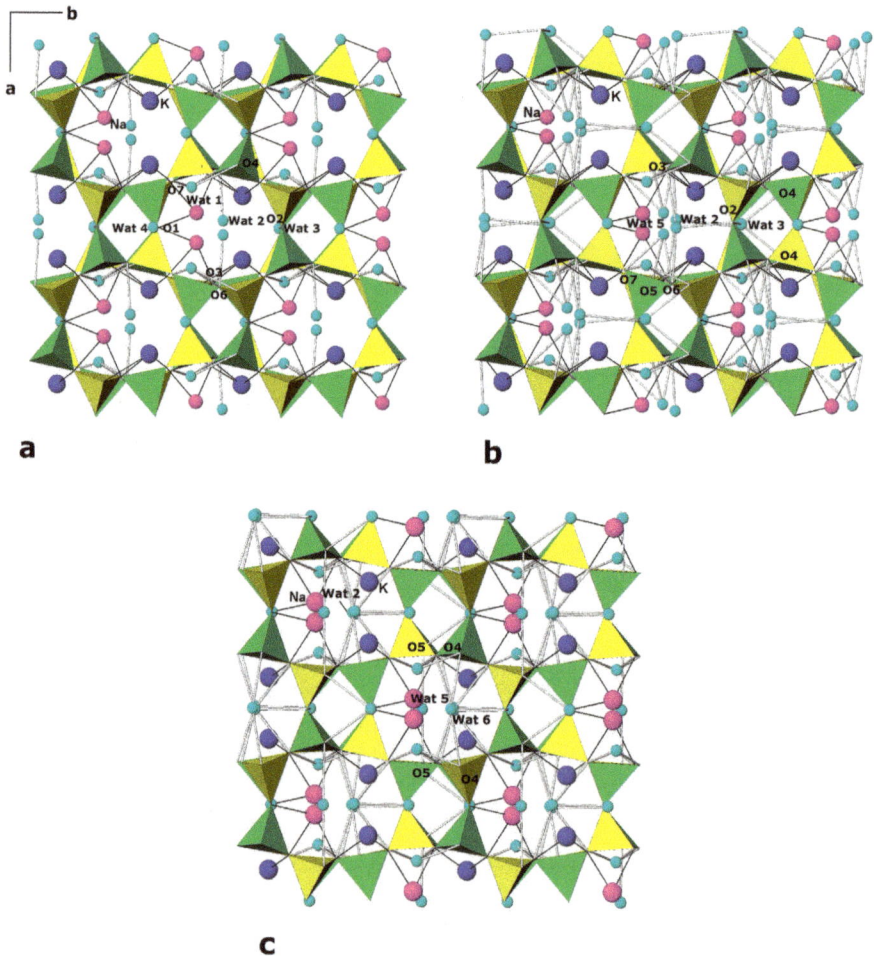

Figure 1. Projection of the amicite structure along the [001] direction at (**a**) P_{amb}, (**b**) 1.25 GPa, and (**c**) 4.71 GPa. The projection of the structure at P_{amb}(rev) is not reported, this being virtually identical to that at P_{amb}. Purple spheres = Na; blue spheres = K; light blue spheres = water; yellow tetrahedra = Si; green tetrahedra = Al.

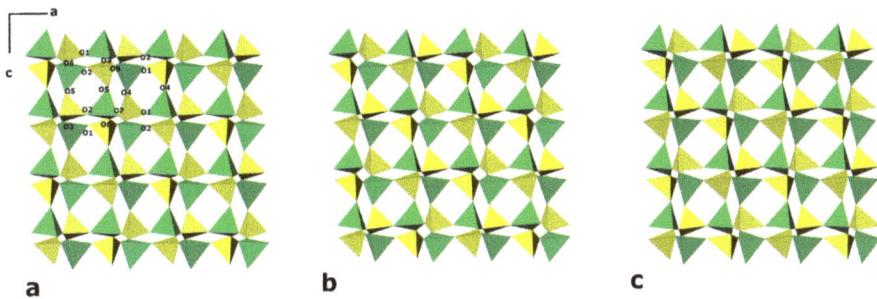

Figure 2. Projection of the amicite structure along the [010] direction at (**a**) P_{amb}, (**b**) 1.25 GPa, and (**c**) 4.71 GPa. The projection of the structure at P_{amb}(rev) is not reported, this being virtually identical to that at P_{amb}.

3. Experimental Methods

In-situ HP XRPD experiments were performed at the SNBL1 (BM01a) beamline at ESRF, using an ETHZ modified Merril-Basset diamond anvil cell (DAC) [29] with flat culets of 600 μm in diameter. Powders were loaded into a pre-indented gasket hole (i.e., a stainless steel foil of 60–80 μm thickness) with 250 μm diameter. The experiments were performed using two different PTM: m.e.w. as nominally penetrating, and s.o. as non-penetrating media, respectively. Pressure was measured before and after data collection at each pressure using the ruby fluorescence method [30] on the non-linear hydrostatic pressure scale [31]. The diffraction data were collected at a wavelength of 0.6825 Å in the Debye–Scherrer geometry on an area detector. One-dimensional diffraction patterns were obtained by integrating the two dimensional images with the program FIT2D [32].

Amicite was compressed up to 8.13(5) GPa in m.e.w. and 8.68(5) GPa in s.o. In the latter case a partial loss of the hydrostatic conditions above 2.8 GPa was observed. In both experiments about 20 images were collected at increasing pressure values. Moreover, some patterns (labeled (rev) in Tables and Figures) were collected upon decompression down to ambient conditions. Figure 3a,b reports selected integrated patterns obtained in m.e.w. and s.o., respectively.

The structural refinements of the data collected in m.e.w. converged successfully up to 4.71(5) GPa. At higher pressure (up to 6.9 GPa) the refinements were still possible, but some framework bond distances and angles produced unreliable values. As a consequence, above 4.71(5) GPa, only the unit-cell parameters were refined by the Rietveld method in the 2°–40° 2θ range.

For amicite in s.o., the low data quality did not allow complete structural refinements. The cell parameters were refined successfully up to 5.48(5) GPa, notwithstanding the previously cited hydrostaticity loss observed above 2.8 GPa.

Rietveld profile fitting was performed using the GSAS package [33] with the EXPGUI [34] interface. The initial structural model is as reported in [28]. The background curve was fitted by a Chebyshev polynomial with 20 coefficients. The pseudo-Voigt profile function proposed by [35] was applied, and the peak intensity cut-off was set to 0.1% of the peak maximum. Soft-restraints were applied to the T–O distances [Si–O = 1.58(2) − 1.62(2); Al–O = 1.72(2) − 1.74(2)] and their weights were gradually decreased after the initial stages of refinement (up to F = 1 in GSAS terminology). The isotropic displacement parameters were constrained in the following way: the same value for all the tetrahedral cations, a second value for all the framework oxygen atoms, a third value for the extraframework cations, and a fourth value for the water molecule oxygen atoms. The unit-cell parameters were allowed to vary in all the refinement cycles. Details of the structural refinements are reported in Table 1.

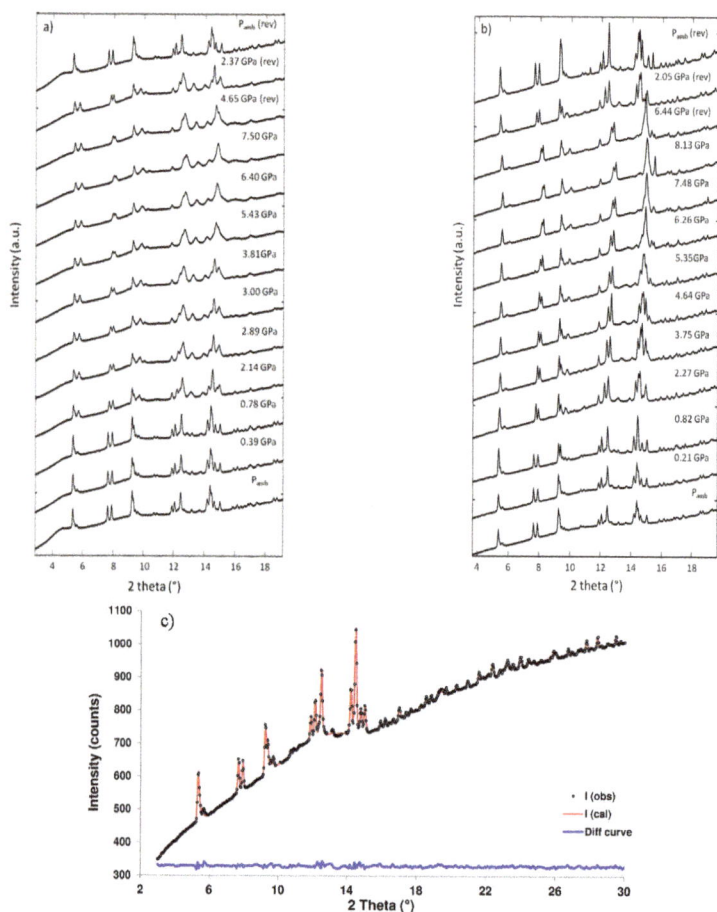

Figure 3. Selected integrated powder patterns, collected in silicone oil (s.o.) (**a**), and (16:3:1) methanol-ethanol-water (m.e.w.) (**b**), reported as a function of pressure. The patterns at the top of the figures were collected during decompression. (**c**) Observed and calculated profiles of the X-ray powder diffraction (XRPD) amicite pattern at 1.25 GPa.

Table 1. Experimental and structural refinement parameters for the X-ray powder diffraction (XRPD) measurements performed on amicite in (16:3:1) methanol:ethanol:water (m.e.w.) at P_{amb}, 1.25 GP, 4.71 GPa, and upon decompression (P_{amb}(rev)).

P (GPa)	P_{amb}	1.25 GPa	4.71 GPa	P_{amb}(rev)
Space Group	I2	I2	I2	I2
a (Å)	10.2324(8)	10.1882(9)	9.8661(5)	10.2296(5)
b (Å)	10.43456(8)	10.443(1)	10.4966(7)	10.4328(6)
c (Å)	9.8987(7)	9.8651(8)	9.6887(5)	9.8931(4)
V (Å^3)	1056.63(2)	1048.7(2)	1002.59(8)	1055.39(8)
$\beta(°)$	88.382(6)	87.49(9)	87.728(8)	88.349(6)
xR_p (%)	0.7	0.7	0.7	1.0
R_{wp} (%)	0.5	0.4	0.4	1.0
R F^2 (%)	11.0	15.3	16.7	11.8
No. of variables	88	94	96	88

4. Results and Discussion

From an inspection of the powder patterns in Figure 3 it is evident that the peak intensities generally decrease and the peak profiles become broader with increasing pressure. These effects could be due to several factors, such as an increase of long-range structural disorder, the presence of texture effects, and in the case of s.o. above 2.8 GPa, a decrease in the hydrostaticity of the PTM. However, HP XRPD data demonstrate that amicite does not undergo complete amorphization up to the highest investigated pressure, and the features characteristic of the pattern collected at ambient conditions are almost completely recovered upon decompression in both experiments.

4.1. Amicite Compressed in Methanol:Ethanol:Water

From P_{amb} to 8.13 GPa, the unit-cell volume reduces by about 9.3%, with the unit-cell axes showing a strongly anisotropic behavior ($\Delta a = -6.0\%$, $\Delta b = +0.6\%$, $\Delta c = -4.1\%$, $\Delta\beta = -0.2\%$) (Table 2 and Figure 4). In particular, the pseudo-tetragonal a and c axes shrink, while the b axis, which is perpendicular to the dense layers, slightly increases. Cell deformation starts above 0.62 GPa and at $P > 3.80$ GPa a slight increase in compressibility is observed (Figure 4d).

Table 2. Unit-cell parameters of natural amicite at the investigated pressures, using (16:3:1) methanol:ethanol:water (m.e.w.) and silicon oil (s.o.) as pressure-transmitting media (PTM).

P (GPa)	a (Å)	b (Å)	c (Å)	V (Å^3)	β (°)
		amicite (m.e.w.)			
P_{amb}	10.2324(8)	10.43456(8)	9.8987(7)	1056.63(2)	88.382(6)
0.04	10.2375(6)	10.4383(6)	9.8949(5)	1056.96(1)	88.343(5)
0.09	10.2378(7)	10.44088(7)	9.8925(6)	1056.96(2)	88.294(5)
0.21	10.241(8)	10.4429(7)	9.886(6)	1056.81	88.170(5)
0.36	10.244(6)	10.44(1)	9.8846(8)	1057(1)	88.03(7)
0.62	10.2447(7)	10.4365(7)	9.8859(6)	1056.4(2)	87.891(6)
0.82	10.2334(7)	10.4289(8)	9.8857(6)	1054.3(2)	87.786(7)
1.25	10.1882(9)	10.443(1)	9.8651(8)	1048.7(2)	87.49(9)
1.67	10.1290(9)	10.471(1)	9.8622(8)	1044.7(2)	87.236(7)
2.22	10.073(1)	10.485(1)	9.853(1)	1039.6(3)	87.395(9)
2.75	10.027(1)	10.498(1)	9.823(1)	1033.1(3)	87.52(1)
3.23	9.9823(6)	10.4975	9.7881(5)	1024.78(9)	87.588(9)
3.80	9.9382(5)	10.5004(6)	9.7507(5)	1016.66(8)	87.633(8)
4.71	9.8661(5)	10.4966(7)	9.6887(5)	1002.59(8)	87.728(8)
5.35	9.820(3)	10.500(4)	9.648(3)	994.1(9)	87.86(3)
6.29	9.747(4)	10.503(5)	9.588(4)	981(1)	87.95(3)
6.71	9.717(4)	10.505(5)	9.568(4)	976(1)	88.08(4)
6.91	9.693(5)	10.502(5)	9.551(4)	972(1)	88.18(4)
7.48	9.657(5)	10.500(6)	9.526(5)	966(1)	88.24(5)
8.13	9.618(6)	10.498(7)	9.496(6)	958(1)	88.24(6)
6.45(rev)	9.732(5)	10.527(5)	9.598(4)	983(1)	88.28(4)
4.42(rev)	9.894(2)	10.528(3)	9.726(2)	1012.4(6)	87.94(2)
2.05(rev)	10.108(1)	10.481(1)	9.884(1)	1046.1(3)	87.44(1)
P_{amb} (rev)	10.2296(5)	10.4328(6)	9.8931(4)	1055.39(8)	88.349(6)
		amicite (s.o.)			
P_{amb}	10.2372(9)	10.4352(9)	9.892(8)	1056.30(2)	88.269(8)
0.39	10.2271(6)	10.4319(7)	9.8686(5)	1052.25(8)	88.071(7)
0.78	10.2133(8)	10.4279(9)	9.8528(7)	1048.70(1)	87.97(1)
1.23	10.1923(9)	10.434(1)	9.8411(7)	1045.90(1)	87.890(1)
1.72	10.160(1)	10.452(2)	9.828(1)	1043(2)	87.8(2)
2.18	10.111(2)	10.468(2)	9.808(2)	1037.4(2)	87.720(3)
2.89	10.026(3)	10.505(4)	9.779(8)	1029.4(4)	88.020(6)
3.35	9.994(4)	10.519(4)	9.77(3)	1026.6(4)	88.16(6)
3.86	9.9440(4)	10.535(5)	9.747(3)	1020.7(5)	88.38(7)
4.27	9.911(4)	10.541(5)	9.729(3)	1016.1(5)	88.55(4)
4.87	9.874(3)	10.537(3)	9.717(2)	1010.5(5)	88.3(4)
5.48	9.816(6)	10.555(6)	9.682(5)	1002.6(2)	88.91(9)
P_{amb} (rev)	10.239(6)	10.432(7)	9.8981(5)	1056.8(2)	88.26(9)

The results of the structural refinements corresponding to four selected pressure values (P_{amb}, 1.25 GPa, 4.71 GPa, P_{amb}(rev)) are reported in Tables 3 and 4 and shown in Figures 1 and 2. The structural variations exhibited by the framework in the range P_{amb}–4.71 GPa regard the shape of both the 4-membered rings, forming the double crankshaft chains, and the 8-membered rings defining the channels along the *a* and *c* axes. In particular: (i) the ellipticity—i.e., the ratio between the longest and shortest oxygen-oxygen distance within the 8-ring window—of the channel running along the *a* axis only slightly decreases, passing from 1.36 at P_{amb} to 1.32 at 4.71 GPa; (ii) the ellipticity of the channel running along the *c* axis considerably increases, passing from 1.39 to 1.54; (iii) the tilted 4-rings of the double crankshaft chains—one defined by the distances O7-O8 and O4-O5, the other by O3-O6 and O1-O2—become a square and a rhombus, respectively; (iv) concerning the two flat 4-rings of the double crankshaft chains, the one defined by the distances O4-O4 and O1-O1 becomes more similar to a square, while the other one defined by O5-O5 and O2-O2 remains almost unchanged.

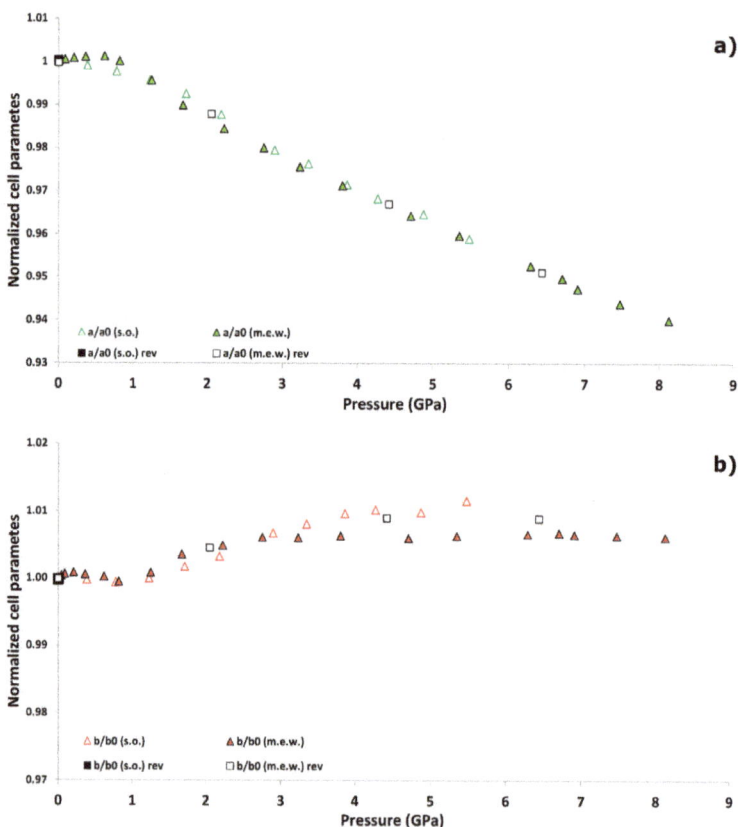

Figure 4. *Cont.*

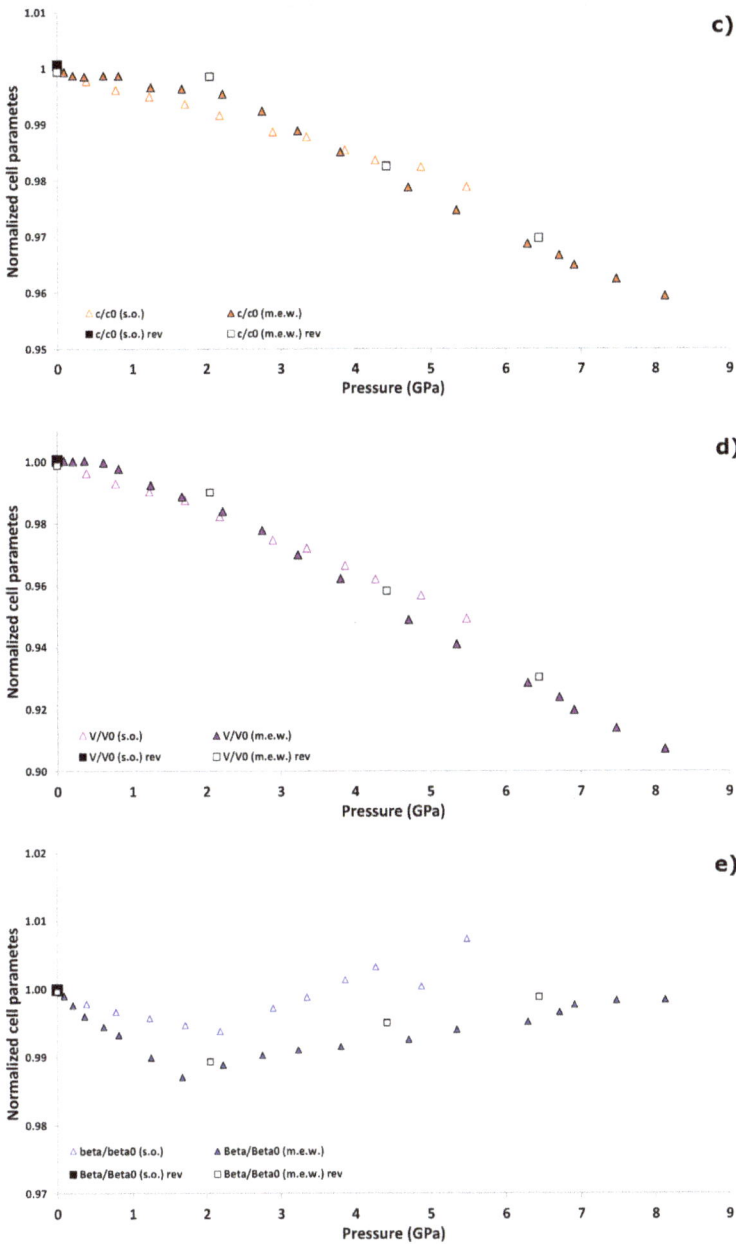

Figure 4. Variation of amicite normalized lattice parameters (**a**) $a/a0$; (**b**) $b/b0$; (**c**) $c/c0$; (**d**) $V/V0$; (**e**) $\beta/\beta0$ as a function of pressure in silicone oil and (16:3:1) methanol-ethanol-water. The errors associated with the cell parameters are smaller than the symbol size.

Table 3. Refined atomic positions, occupancy factors, and displacement parameters of amicite at P_{amb}, 1.25 GPa, 4.71 GPa, and upon decompression ($P_{amb}(rev)$) in m.e.w.

	x/a	y/b	z/c	Occ	Uiso (*100)
			P_{amb}		
K	0.324(1)	−0.014(2)	0.960(1)	1	0.91(3)
Na	0.438(1)	0.241(2)	0.661(1)	1	0.91(3)
Al1	0.153(2)	0.241(3)	0.150(2)	1	0.6(3)
Al2	0.152(2)	−0.005(3)	0.658(2)	1	0.6(3)
Si1	0.156(2)	−0.023(3)	0.327(2)	1	0.6(3)
Si2	0.159(2)	0.255(3)	0.820(2)	1	0.6(3)
O1	0.006(2)	−0.055(4)	0.303(4)	1	1.5(3)
O2	−0.006(2)	0.273(5)	0.205(3)	1	1.5(3)
O3	0.206(3)	0.135(3)	0.730(4)	1	1.5(3)
O4	0.176(3)	0.008(5)	0.484(2)	1	1.5(3)
O5	0.182(3)	0.226(5)	0.977(2)	1	1.5(3)
O6	0.187(4)	0.107(3)	0.247(4)	1	1.5(3)
O7	0.255(4)	0.356(3)	0.215(3)	1	1.5(3)
O8	0.765(3)	0.386(3)	0.216(4)	1	1.5(3)
Wat1	0.307(4)	0.267(6)	0.475(5)	0.75(3)	0.2(6)
Wat2	0.456(3)	0.090(3)	0.243(3)	1	0.2(6)
Wat3	0	0.300(6)	0.5	1	0.2(6)
Wat4	0.5	0.42(4)	0.5	0.12(3)	0.2(6)
			1.25 GPa		
K	0.331(1)	-0.001(2)	0.964(1)	1	4.5(4)
Na	0.449(2)	0.276(3)	0.672(2)	1	4.5(4)
Al1	0.143(1)	0.249(2)	0.147(2)	1	0.4(2)
Al2	0.164(2)	0.007(2)	0.659(2)	1	0.4(2)
Si1	0.151(1)	−0.010(2)	0.331(2)	1	0.4(2)
Si2	0.159(1)	0.265(3)	0.822(2)	1	0.4(2)
O1	−0.003(1)	−0.031(3)	0.301(3)	1	0.7(3)
O2	−0.008(1)	0.307(3)	0.201(3)	1	0.7(3)
O3	0.218(2)	0.146(3)	0.730(4)	1	0.7(3)
O4	0.182(3)	0.028(4)	0.486(1)	1	0.7(3)
O5	0.183(3)	0.230(4)	0.977(2)	1	0.7(3)
O6	0.183(3)	0.117(3)	0.241(3)	1	0.7(3)
O7	0.256(3)	0.365(3)	0.185(3)	1	0.7(3)
O8	0.756(3)	0.389(3)	0.230(2)	1	0.7(3)
Wat1	0.283(2)	0.257(5)	0.467(3)	1	1.7(5)
Wat2	0.480(3)	0.105(4)	0.277(3)	1	1.7(5)
Wat3	0	0.259(6)	0.5	1	1.7(5)
Wat4	0.5	0.459(7)	0.5	0.74(3)	1.7(5)
Wat5	0.570(6)	0.172(5)	0.964(4)	0.40(3)	1.7(5)
			4.71 GPa		
K	0.340(1)	0.006(2)	0.965(2)	1	5.1(5)
Na	0.448(2)	0.276(3)	0.672(2)	1	5.1(5)
Al1	0.164(2)	0.259(2)	0.154(2)	1	2.1(3)
Al2	0.166(2)	0.024(3)	0.657(2)	1	2.1(3)
Si1	0.161(2)	0.009(2)	0.32(2)	1	2.1(3)
Si2	0.152(2)	0.275(3)	0.822(2)	1	2.1(3)
O1	0.001(2)	−0.009(4)	0.303(3)	1	2.4(4)
O2	0.004(2)	0.316(4)	0.199(4)	1	2.4(4)
O3	0.220(3)	0.167(3)	0.723(3)	1	2.4(4)
O4	0.206(3)	0.044(4)	0.483(2)	1	2.4(4)
O5	0.174(3)	0.229(3)	0.980(2)	1	2.4(4)
O6	0.216(3)	0.143(3)	0.265(3)	1	2.4(4)
O7	0.225(3)	0.403(3)	0.209(3)	1	2.4(4)
O8	0.766(3)	0.404(3)	0.218(4)	1	2.4(4)
Wat1	0.290(2)	0.259(5)	0.472(4)	1	2.5(5)
Wat2	0.494(3)	0.078(3)	0.247(3)	1	2.5(5)
Wat3	0	0.301(1)	0.5	0.80(3)	2.5(5)
Wat4	0.5	0.494(7)	0.5	1	2.5(5)
Wat5	0.5	0.234(7)	0	1	2.5(5)
Wat6	0	0.590(10)	0	0.48(3)	2.5(5)

Table 3. *Cont.*

	x/a	y/b	z/c	Occ	Uiso (*100)
		P_{amb}(rev)			
K	0.335(1)	0.002(2)	0.967(1)	1	3.0(4)
Na	0.446(2)	0.270(3)	0.671(2)	1	3.0(4)
Al1	0.143(1)	0.256(2)	0.154(2)	1	0.3(2)
Al2	0.164(2)	0.017(3)	0.651(2)	1	0.3(2)
Si1	0.153(1)	−0.012(2)	0.323(2)	1	0.3(2)
Si2	0.166(1)	0.269(2)	0.824(2)	1	0.3(2)
O1	−0.004(1)	−0.017(3)	0.303(3)	1	0.4(4)
O2	−0.013(1)	0.304(3)	0.198(3)	1	0.4(4)
O3	0.220(3)	0.151(3)	0.732(4)	1	0.4(4)
O4	0.186(3)	0.027(4)	0.478(2)	1	0.4(4)
O5	0.174(4)	0.233(4)	0.983(2)	1	0.4(4)
O6	0.183(3)	0.121(3)	0.244(4)	1	0.4(4)
O7	0.244(3)	0.374(3)	0.213(3)	1	0.4(4)
O8	0.767(3)	0.405(3)	0.218(3)	1	0.4(4)
Wat1	0.278(2)	0.271(5)	0.473(3)	0.92(2)	0.2(7)
Wat2	0.449(3)	0.104(2)	0.241(3)	1	0.2(7)
Wat3	0	0.329(5)	0.5	1	0.2(7)
Wat4	0.5	0.457(3)	0.5	0.16(3)	0.2(7)

Table 4. Framework and extraframework distances (<3.20 Å) for amicite at P_{amb}, 1.25 GPa, 4.71 GPa, and upon decompression (P_{amb}(rev)) in m.e.w.

		P_{amb}	1.25 GPa	4.71 GPa	P_{amb}(rev)
Al1-	O2	1.732(3)	1.721(2)	1.721(3)	1.721(3)
	O5	1.738(3)	1.722(2)	1.721(3)	1.727(3)
	O6	1.732(3)	1.720(2)	1.719(3)	1.720(3)
	O7	1.732(3)	1.720(2)	1.719(3)	1.721(3)
Al2-	O1	1.733(3)	1.721(2)	1.720(3)	1.723(3)
	O3	1.731(3)	1.721(2)	1.720(3)	1.721(3)
	O4	1.735(3)	1.721(2)	1.720(3)	1.724(3)
	O8	1.731(3)	1.720(2)	1.720(3)	1.719(3)
Si1-	O1	1.603(3)	1.620(2)	1.620(3)	1.622(3)
	O4	1.606(3)	1.620(2)	1.620(3)	1.624(3)
	O6	1.601(3)	1.620(2)	1.619(3)	1.620(3)
	O7	1.601(3)	1.620(2)	1.619(3)	1.620(3)
Si2-	O2	1.602(3)	1.6202(2)	1.620(3)	1.622(3)
	O3	1.601(3)	1.6203(2)	1.620(3)	1.621(3)
	O5	1.608(3)	1.6211(2)	1.621(3)	1.626(3)
	O8	1.600(3)	1.6196(2)	1.620(3)	1.620(3)
K-	O3	3.07(1)	3.06(2)	3.11(2)	3.06(1)
	O5	2.89(2)	2.78(2)	2.82(1)	2.92(2)
	O8	2.75(1)	2.76(2)	2.77(1)	2.78(1)
	Wat1	2.71(4)	2.86(4)	2.94(1)	2.73(4)
	Wat2	3.19(2)	3.19(3)	2.69(2)	3.16(3)
	Wat3	2.67(4)	3.07(5)	2.69(1)	3.16(2)
	Wat5		3.20(7)	2.85(1)	2.50(4)
	Wat5		2.28(8)	2.85(1)	
Na-	O1	2.65(2)	2.38(2)	2.74(1)	2.63(3)
	O3	2.68(1)	2.77(2)	2.52(1)	2.68(2)
	O8	2.822)	2.52(2)	2.68(2)	2.79(2)
	Wat1	2.33(5)	2.70(4)	2.50(2)	2.65(3)
	Wat1	2.92(3)	3.00(2)	2.98(2)	3.13(3)
	Wat2	2.15(3)	2.10(5)	2.29(1)	2.23(3)
	Wat4	2.52(3)	2.59(5)	2.86(1)	2.62(2)
	Wat5				
	Wat6			2.54(1)	

Table 4. *Cont.*

		P_{amb}	1.25 GPa	4.71 GPa	P_{amb}(rev)
Wat1-	K		2.86(4)	2.94(1)	2.73(4)
	Na	2.33(5)	2.70(4)	2.50(2)	2.65(3)
	Na	2.92(3)	3.00(2)	2.98(2)	3.13(3)
	O3	3.02(4)	2.88(3)	2.63(1)	2.90(3)
	O4	3.01(6)	2.61(5)	2.37(1)	2.70(5)
	O6	3.09(6)	2.82(4)	2.48(1)	2.93(4)
	O7	2.79(5)	3.03(3)	3.05(1)	2.82(3)
	Wat2		3.12(5)		
	Wat3	3.16(4)	2.89(2)	2.84(1)	2.91(2)
	Wat4	2.55(3)	3.08(6)		3.01(2)
	Wat6			2.75(1)	
Wat2-	K	3.191(2)	3.19(3)	2.69(2)	3.16(3)
	K				3.16(2)
	Na	2.152(3)	2.10(5)	2.29(1)	2.23(3)
	O2		3.13(4)	2.78(2)	
	O3		3.11(4)	3.02(1)	
	O6	2.752(3)	3.07(4)	2.77(1)	2.73(3)
	O7			2.83(1)	3.14(2)
	O8			3.081)	
	Wat1		3.12(4)		
	Wat5		2.556(3)	2.911)	
	Wat6			2.42(1)	
Wat3-	K(x2)	2.67(4)	3.07(5)	2.69(1)	2.50(4)
	O2(x2)	2.939(6)	3.054(9)	2.911)	3.009(4)
	O4 (x2)		3.02(5)		
	Wat1(x2)	3.16(4)	2.89(2)	2.84(1)	2.911(2)
Wat4-	Na(x2)	2.52(3)	2.59(5)	2.86(1)	2.62(2)
	O1(x2)	3.01(4)	2.973(7)	2.93(1)	3.008(3)
	Wat1 (x2)		3.08(6)	2.96(1)	3.01(2)
Wat5-	K		3.03(6)	2.85(1)	
	K		2.19(8)	2.85(1)	
	Na		3.03(8)		
	O5		2.67(8)		
	O6		3.20(6)		
	O7		3.01(7)		
	Wat2		2.556(3)	2.91(1)	
	Wat2			2.91(1)	
Wat6-	Na (x2)			2.54(1)	
	O4 (x2)			2.97(2)	
	Wat1 (x2)			2.75(1)	
	Wat2 (x2)			2.42(1)	

Beyond these framework deformations, the most remarkable effect induced on amicite by compression in m.e.w. is the penetration of additional water molecules from the aqueous PTM into the pores (Figure 5a,b). Already at 0.04 GPa, the W1 site, originally partially occupied, fills up and W4 increases its occupancy factor. At 0.35 GPa a new water site (W5) appears near the two-fold axis parallel to b. Its occupancy factor subsequently tends to increase up to the maximum and the water molecule moves to the two-fold axis (Figures 1 and 5a and Table 3). Another water site (W6) appears at 3.2 GPa with an occupancy factor of about 0.5, which remains unvaried up to 4.71 GPa. Both sites are close to the center of the 8-membered channel parallel to [001]. In the investigated P range the total number of water molecules increases from 9.24 to 14.58 (see Table 3 and Figure 5b). The cations and the original water molecules undergo only slight positional changes and the new water sites W5 and W6 enter into the coordination sphere of K and Na, respectively.

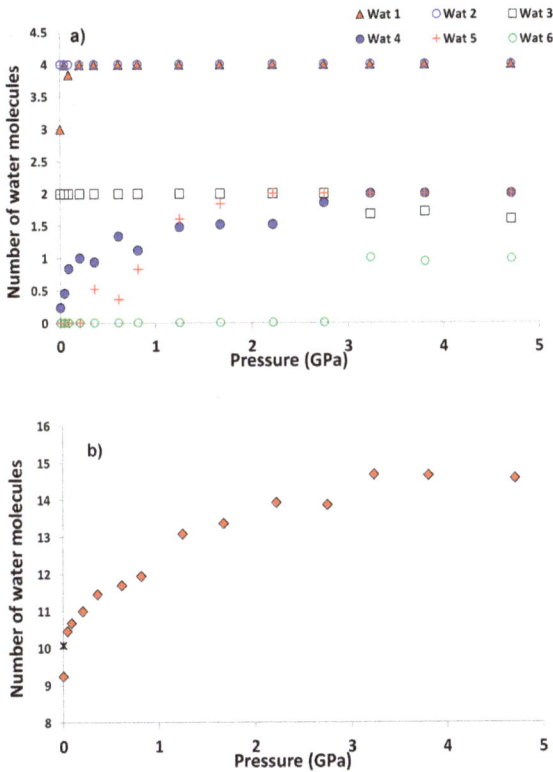

Figure 5. (a) Number of water molecules in the Wat1-Wat6 extraframework sites in amicite compressed in m.e.w. as a function of pressure; (b) total number of water molecules as a function of P. The cross indicates the number of water molecules after pressure release.

In Figure 4d two slope changes are visible in the P-V diagram at P > 0.65 and P > 3.2 GPa, and these can be considered as strictly related to the water penetration. In fact, the rather strong bonds W5-O5 and W6-O4 between the new water sites and the framework, and the decrease in the distances between W1 and the oxygen atoms O3, O4, O6 (Table 4)—along with the W1 filling—contribute to the *a* axis contraction, which is the main effect responsible of amicite compression behavior.

The structure refinement performed after P release to ambient conditions (P_{amb}(rev)), shows that the unit cell parameters (Table 2) and all the structural features (Tables 3 and 4) recover their original values. The W5 and W6 sites disappear and only W1 maintains a higher occupancy factor compared to the original amicite [28]. Overall, less than one additional water molecule of the PTM remains in the structure.

4.2. Amicite Compressed in Silicone Oil

Table 2 and Figure 4 show the P-dependence of the lattice parameters of amicite in s.o. It can be seen that the pseudotetragonal *a* and *c* axes initially decrease by approximately the same percentage, while the *b* axis slightly increases. The *a* and *c* parameters decrease quite regularly, with a slight slope increase between 2 and 3 GPa, particularly evident for the *a* parameter. In the range P_{amb}–5.43 GPa, the overall cell parameter variations are: $\Delta a = -4.1\%$, $\Delta b = +1.1\%$, $\Delta c = -2.1\%$, $\Delta \beta = -0.7\%$, while the cell volume decreases by approximately 5.1%. Again in s.o. the P-induced effects on the amicite unit cell are reversible upon decompression, as shown in Table 2 and Figure 4.

5. Comparison between Amicite Compressibility in Aqueous Medium and Silicone Oil

The main difference between amicite HP behavior in m.e.w. and s.o. is the higher compressibility in the aqueous medium (see Figure 4 and Table 2). This is clear by comparing the unit cell volume decrease in the two PTM at similar pressure values: 5.1% at 5.43 GPa in s.o., and 5.9% at 5.35 GPa in m.e.w. This effect is anomalous compared to what is generally observed for zeolites, when water penetration provides a support against the effects of pressure (e.g., see a review in Table 4 in [36]) [12,13,37,38]. Although detailed structural data for the ramp in s.o. are lacking, this result can be ascribed to formation, during compression in m.e.w., of rather strong bonds between the additional water molecules and the framework oxygen atoms, which contribute to the shrinkage of the *a* parameter. In particular, the distance O3–O6, which is parallel to the *a* axis and corresponds to the shortest diameter of the 8-ring perpendicular to *c*, undergoes a 10% reduction passing from 6.20 Å at P_{amb} to 5.58 Å at 4.71 GPa (see Figure 1).

6. Compressibility Behavior of Microporous Materials with GIS Topology

A number of microporous materials with GIS topology have been investigated under HP. Among the natural zeolites, amicite, the K-Na member of the GIS family, can be compared to gismondine [11,22], the Ca member, producing the following observations:

1. Compression of gismondine in both m.e.w. and s.o. favors the tetragonalization of the unit cell; in amicite the *a* and *c* axes also tend to become more similar at HP, but the beta angle does not substantially change;
2. Gismondine compressed in m.e.w. undergoes a transition to a triclinic phase at about 3 GPa; the original symmetry of amicite, by contrast, is maintained in both the experiments;
3. The HP framework deformation mechanism is the same in the two zeolites, essentially being driven by the distortion of the "double crankshaft" chains and the consequent change in the 8-ring channel shape;
4. Amicite's compressibility increases at HP both in m.e.w. and s.o.; by contrast, gismondine's compressibility in s.o. slightly decreases while in m.e.w. it remains constant;
5. PIH occurs in both amicite and in gismondine compressed in m.e.w. However, it induces different reorganizations in the water molecule systems: in amicite there is both the filling of partially occupied sites and the appearance of two new water sites; in gismondine four partially occupied water sites reduce to only two fully occupied sites, giving rise to a more ordered water system;
6. In amicite 5.34 water molecules enter the zeolite porosities when compressed in m.e.w., while in gismondine only one additional molecule penetrates. This result can be explained by the higher channel stuffing of gismondine at P_{amb} compared to amicite;
7. Both amicite and gismondine are more compressible in m.e.w. than in s.o., but for different reasons. In gismondine this effect has been justified by the re-organization of the water molecule system, which leaves a larger free volume inside the pores compared to the phase compressed in s.o. In amicite the higher compressibility at HP results from the strong bonds between framework oxygen atoms and the new water molecules;
8. Overall, gismondine is more compressible than amicite, both in m.e.w. and in s.o. Comparing the unit cell volume decrease of the two phases at a similar pressure value—about 5.5 GPa—we find $\Delta V = -7.5\%$ and -6.4% for gismondine in m.e.w. and s.o., respectively, while for amicite these values are -5.9% and -5.1%, respectively. The presence of the large potassium cations and the higher number of extraframework sites after PIH in amicite compared to gismondine probably contribute to better supporting the amicite structure.

Lee and co-workers [23] studied the compressibility in m.e.w. of the potassium gallo silicate K-GaSi-GIS with GIS framework type (ideal formula $K_{5.76}Ga_{5.76}Si_{10.24}O_{32}\cdot9.9H_2O$, s.g. $I4_1/a$). When the results of this study are compared with those obtained from the natural phases, the following observations can be made:

(a) The main feature of the P-induced evolution of cell parameters of K-GaSi-GIS is the noticeable squashing of the *c* axis, which is perpendicular to the dense plane and corresponds to the *b* axis of gismondine and amicite. This response to hydrostatic pressure corresponds to a gradual flattening of the double crankshaft chains and a reduction in the ellipticity of the 8-ring windows. The different behavior compared to amicite and gismondine, where the *b* axis slightly increases or remains almost unvaried, could be explained by the lower channel stuffing of the K-GaSi-GIS phase related to the high Si/Ga ratio;

(b) In K-GaSi-GIS a PIH effect is again observed, with the penetration of about two water molecules at P < 1 GPa, but in this case the overhydration induces a disordering of the K-water system along the channels.

The potassium alumino germanate K-AlGe-GIS with GIS topology (ideal formula $K_8Al_8Ge_8O_{32}\cdot8H_2O$, s.g. $I2/a$) was studied under HP by Jang et al. [24]. Its structure is similar to amicite for the ordered distribution of the tetrahedral cations and the same number of extraframework cations. However, there are eight instead of 10 water molecules in the synthetic phase. The variation in the unit cell parameters was determined in m.e.w. up to 3.22 GPa, but no structural refinements were reported, so no hypotheses were made concerning a possible PIH. The compressibility is anisotropic with a decrease in the *a* and *c* axes, parallel to the channels, of 1.3% and 1.0%, respectively, while the *b* parameter, perpendicular to the channels and the double crankshaft chains, decreased by 5.4% resulting in an almost linear volume contraction of 7.5%. The large *b* variation is strictly related to a flattening of the double crankshaft chains under P, as already observed for K-GaSi-GIS. Again in this case, the different compressibility behavior compared to amicite and gismondine can be explained by the lower water content and consequent channel stuffing.

7. Conclusions

The high-pressure behavior of amicite, a GIS framework type zeolite, was investigated and the strong influence on compressibility of the chemical composition of both the framework and extraframework species was also confirmed for this variety. In particular, the study confirms that the compressibility of microporous materials is not simply related to their framework density and topology, but is also greatly affected by the type, amount, and location of the extra-framework species.

The HP framework deformation mechanism is the same in all the phases with GIS topology and is essentially driven by the distortion of the "double crankshaft" chains and the consequent shape change of the 8-ring channels. However, the degree of compressibility varies due to the different chemical compositions. In these zeolites the pressure-induced penetration of water molecules does not induce a unit cell volume expansion, and in the natural phases, when the structure after P release was determined, the overhydration effects are reversible following the return to ambient conditions.

Acknowledgments: This work was supported by the Italian MIUR, within the frame of the following projects: PRIN2015 "ZAPPING" High-pressure nano-confinement in Zeolites: the Mineral Science know-how APPlied to engineerING of innovative materials for technological and environmental applications" (2015HK93L7), and FIRB, Futuro in Ricerca "Impose Pressure and Change Technology" (RBFR12CLQD). Vladimir Dmitriev and the ESRF BM01 beamline staff are acknowledged for their technical support during the high-pressure XRPD experiments.

Author Contributions: Rossella Arletti, Simona Quartieri, and Giovanna Vezzalini conceived and designed the experiments; Rossella Arletti, Carlotta Giacobbe, and Giovanna Vezzalini performed the experiments; Rossella Arletti and Carlotta Giacobbe analyzed the data; Rossella Arletti, Simona Quartieri, and Giovanna Vezzalini wrote the paper.

Conflicts of Interest: The authors declare no conflict of interest.

References

1. Gillet, P.; Malezieux, J.M.; Itie, J.P. Phase changes and amorphization of zeolites at high pressure: The case of scolecite and mesolite. *Am. Mineral.* **1996**, *81*, 651–657. [CrossRef]
2. Huang, Y.; Havenga, E.A. Why do zeolites with LTA structure undergo reversible amorphization under pressure? *Chem. Phys. Lett.* **2001**, *345*, 65–71. [CrossRef]
3. Rutter, M.D.; Uchida, T.; Secco, R.A.; Huang, Y.; Wang, Y. Investigation of pressure-induced amorphization in hydrated zeolite Li-A and Na-A using synchrotron X-ray diffraction. *J. Phys. Chem. Solids* **2001**, *62*, 599–606. [CrossRef]
4. Greaves, G.N.; Meneau, F.; Sapelkin, A.; Colyer, L.M.; Gwynn, I.A.; Wade, S.; Sankar, G. The rheology of collapsing zeolites amorphized by temperature and pressure. *Nat. Mater.* **2003**, *2*, 622–629. [CrossRef] [PubMed]
5. Gulìn-González, J.; Suffritti, G.B. Amorphization of calcined LTA zeolites at high pressure: A computational study. *Microporous Mesoporous Mater.* **2004**, *69*, 127–134. [CrossRef]
6. Goryainov, S.V. Pressure-induced amorphization of $Na_2Al_2Si_3O_{10}\cdot2H_2O$ and $KAlSi_2O_6$ zeolites. *Phys. Status Solidi* **2005**, *202*, R25–R27. [CrossRef]
7. Secco, R.A.; Huang, Y. Pressure-induced disorder in hydrated Na-A zeolite. *J. Phys. Chem. Solids* **1999**, *60*, 999–1002. [CrossRef]
8. Rutter, M.D.; Secco, R.A.; Huang, Y. Ionic conduction in hydrated zeolite Li-, Na- and K-A at high pressures. *Chem. Phys. Letter* **2000**, *331*, 189–195. [CrossRef]
9. Lee, Y.; Vogt, T.; Hriljac, J.A.; Parise, J.B.; Artioli, G. Pressure-induced volume expansion of zeolites in the natrolite family. *J. Am. Chem. Soc.* **2002**, *124*, 5466–5475. [CrossRef] [PubMed]
10. Lee, Y.; Hriljac, J.A.; Vogt, T. Pressure-induced migration of zeolitic water in laumontite. *Phys. Chem. Miner.* **2004**, *31*, 421–428. [CrossRef]
11. Ori, S.; Quartieri, S.; Vezzalini, G.; Dmitriev, V. Pressure-induced over-hydration and water ordering in gismondine: A synchrotron powder diffraction study. *Am. Mineral.* **2008**, *93*, 1393–1403. [CrossRef]
12. Arletti, R.; Quartieri, S.; Vezzalini, G. Elastic behavior of zeolite boggsite in silicone oil and aqueous medium: A case of high-pressure-induced over-hydration. *Am. Mineral.* **2010**, *95*, 1247–1256. [CrossRef]
13. Quartieri, S.; Montagna, G.; Arletti, R.; Vezzalini, G. Elastic behavior of MFI-type zeolites: Compressibility of H-ZSM-5 in penetrating and non-penetrating media. *J. Solid State Chem.* **2011**, *184*, 1505–1516. [CrossRef]
14. Lotti, P.; Gatta, G.D.; Comboni, D.; Merlini, M.; Pastero, L.; Hanfland, M. AlPO4-5 zeolite at high pressure: Crystal–fluid interaction and elastic behavior. *Microporous Mesoporous Mater.* **2016**, *228*, 158–167. [CrossRef]
15. Lee, Y.; Hriljac, J.A.; Vogt, T. Pressure-induced argon insertion into an auxetic small pore zeolite. *J. Phys. Chem. C* **2010**, *114*, 6922–6927. [CrossRef]
16. Seoung, D.; Lee, Y.; Cynn, H.; Park, C.; Choi, K.Y.; Blom, D.A.; Evans, W.J.; Kao, C.C.; Vogt, T.; Lee, Y. Irreversible xenon insertion into a small-pore zeolite at moderate pressures and temperatures. *Nat. Chem.* **2014**, *6*, 835–839. [CrossRef] [PubMed]
17. Lee, Y.; Liu, D.; Seoung, D.; Liu, Z.; Kao, C.C.; Vogt, T. Pressure- and heat-induced insertion of CO_2 into an auxetic small-pore zeolite. *J. Am. Chem. Soc.* **2011**, *133*, 1674–1677. [CrossRef] [PubMed]
18. Santoro, M.; Scelta, D.; Dziubek, K.; Ceppatelli, M.; Gorelli, F.A.; Bini, R.; Garbarino, G.; Thibaud, J.M.; Di Renzo, F.; Cambon, O.; et al. Synthesis of 1D polymer/zeolite nanocomposites under high pressure. *Chem. Mater.* **2016**, *28*, 4065–4071. [CrossRef]
19. Gatta, G.D.; Lee, Y. Zeolites at high pressure: A review. *Mineral. Mag.* **2014**, *78*, 267–291. [CrossRef]
20. Baerlocher, C.; McCusker, L.B.; Olson, D.H. *Atlas of Zeolite Framework Types*, 6th ed.; Elsevier: Amsterdam, The Netherlands, 2007.
21. Vezzalini, G.; Quartieri, S.; Alberti, A. Structural modifications induced by dehydration in the zeolite gismondine. *Zeolites* **1993**, *13*, 34–42. [CrossRef]
22. Betti, C.; Fois, E.; Mazzucato, E.; Medici, C.; Quartieri, S.; Tabacchi, G.; Vezzalini, G.; Dmitriev, V. Gismondine under HP: Deformation mechanism and re-organization of the extra-framework species. *Microporous Mesoporous Mater.* **2007**, *103*, 190–209. [CrossRef]
23. Lee, Y.; Kim, S.J.; Kao, C.C.; Vogt, T. Pressure-induced hydration and order-disorder transition in a synthetic potassium gallosilicate zeolite with gismondine topology. *J. Am. Chem. Soc.* **2008**, *130*, 2842–2850. [CrossRef] [PubMed]

24. Jang, Y.N.; Kao, C.C.; Vogt, T.; Lee, Y. Anisotropic compression of a synthetic potassium aluminogermanate zeolite with gismondine topology. *J. Solid State Chem.* **2010**, *183*, 2305–2308. [CrossRef]
25. Rocha, J.; Lin, Z. Micro and mesoporous mineral phases. *Rev. Mineral. Geochem.* **2005**, *57*, 173–201. [CrossRef]
26. Danisi, R.M.; Armbruster, T.; Arletti, R.; Gatta, G.D.; Vezzalini, G.; Quartieri, S.; Dmitriev, V. Elastic behavior and pressure-induced structural modifications of the microporous Ca(VO)Si$_4$O$_{10}$·4H$_2$O dimorphs cavansite and pentagonite. *Microporous Mesoporous Mater.* **2015**, *204*, 257–268. [CrossRef]
27. Coombs, D.S.; Alberti, A.; Armbruster, T.; Artioli, G.; Colella, C.; Galli, E.; Grice, J.D.; Liebau, F.; Mandarino, J.A.; Minato, H.; et al. Recommended nomenclature for zeolite minerals: Report of the subcommittee on zeolites of the international mineralogical association, commission on new minerals and minerals names. *Can. Mineral.* **1997**, *35*, 1571–1606.
28. Alberti, A.; Vezzalini, G. The crystal structure of amicite, a zeolite. *Acta Cryst.* **1979**, *B35*, 2866–2869. [CrossRef]
29. Miletich, R.; Allan, D.R.; Kush, W.F. High-temperature and high-pressure crystal chemistry. *Rev. Mineral. Geochem.* **2000**, *41*, 445–519. [CrossRef]
30. Forman, R.A.; Piermarini, G.J.; Barnett, J.D.; Block, S. Pressure measurement made by the utilization of ruby sharp-line luminescence. *Science* **1972**, *176*, 4673–4676. [CrossRef] [PubMed]
31. Mao, H.K.; Xu, J.; Bell, P.M. Calibration of the ruby pressure gauge to 800 kbar under quasi-hydrostatic conditions. *J. Geophys. Res.* **1986**, *9*, 4673–4676. [CrossRef]
32. Hammersley, A.P.; Svensson, S.O.; Hanfland, M.; Fitch, A.N.; Häusermann, D. Two dimensional detector software: From real detector to idealized image or two-theta scan. *High Press. Res.* **1996**, *14*, 235–248. [CrossRef]
33. Larson, A.C.; von Dreele, R.B. *GSAS-General Structure Analysis System*; Report LAUR 86-748; Los Alamos National Laboratory: Los Alamos, NM, USA, 1996.
34. Toby, B.H. EXPGUI, a graphical user interface for GSAS. *J. Appl. Cryst.* **2001**, *34*, 210–213. [CrossRef]
35. Thomson, P.; Cox, D.E.; Hastings, J.B. Rietveld refinement of Debye–Scherrer synchrotron X-ray data from Al$_2$O$_3$. *J. Appl. Cryst.* **1987**, *20*, 79–83. [CrossRef]
36. Colligan, M.; Forster, P.M.; Cheetham, A.K.; Lee, Y.; Vogt, T.; Hriljac, J.A. Synchrotron X-ray powder diffraction and computational investigation of purely siliceous zeolite Y under pressure. *J. Am. Chem. Soc.* **2004**, *126*, 12015–12022. [CrossRef] [PubMed]
37. Likhacheva, A.Y.; Seryotkin, Y.V.; Manakov, A.Y.; Goryainov, S.V.; Ancharov, A.I.; Sheromov, M.A. Anomalous compression of scolecite and thomsonite in aqueous medium to 2 GPa. *High Press. Res.* **2006**, *26*, 449–453. [CrossRef]
38. Likhacheva, A.Y.; Seryotkin, Y.V.; Manakov, A.Y.; Goryainov, S.V.; Ancharov, A.I.; Sheromov, M.A. Pressure-induced over-hydration of thomsonite: A synchrotron powder diffraction study. *Am. Mineral.* **2007**, *92*, 1610–1615. [CrossRef]

minerals

MDPI

Article

Effect of Silica Alumina Ratio and Thermal Treatment of Beta Zeolites on the Adsorption of Toluene from Aqueous Solutions

Elena Sarti [1], Tatiana Chenet [1], Luisa Pasti [1,*], Alberto Cavazzini [1], Elisa Rodeghero [2] and Annalisa Martucci [2]

[1] Department of Chemical and Pharmaceutical Sciences, University of Ferrara, Via Fossato di Mortara 17, 44121 Ferrara, Italy; elena.sarti@unife.it (E.S.); tatiana.chenet@unife.it (T.C.); alberto.cavazzini@unife.it (A.C.)

[2] Department of Physics and Earth Sciences, University of Ferrara, Via Saragat 1, 44122 Ferrara, Italy; elisa.rodeghero@unife.it (E.R.); annalisa.martucci@unife.it (A.M.)

* Correspondence: luisa.pasti@unife.it; Tel.: +39-0532-455-346

Academic Editor: Huifang Xu
Received: 23 December 2016; Accepted: 10 February 2017; Published: 15 February 2017

Abstract: The adsorption of toluene from aqueous solutions onto hydrophobic zeolites was studied by combining chromatographic, thermal and structural techniques. Three beta zeolites (notated BEAs, since they belong to BEA framework type), with different SiO_2/Al_2O_3 ratios (i.e., 25, 38 and 360), before and after calcination, were tested as adsorbents of toluene from aqueous media. This was performed by measuring the adsorbed quantities of toluene onto zeolites in a wide concentration range of solute. The adsorption data were fitted with isotherms whose models are based on surface heterogeneity of the adsorbent, according to the defective structure of beta zeolites. The thermal treatment considerably increases the adsorption of toluene, in the low concentration range, on all BEAs, probably due to surface and structural modifications induced by calcination. Among the calcined BEAs, the most hydrophobic zeolite (i.e., that with SiO_2/Al_2O_3 ratio of 360) showed the highest binding constant, probably due to its high affinity for an organophilic solute such as toluene. The high sorption capacity was confirmed by thermogravimetric analyses on BEAs, before and after saturation with toluene.

Keywords: adsorption; zeolites; beta; toluene

1. Introduction

In recent years public concern has been rapidly grown regarding water pollution phenomena. Petroleum hydrocarbons represent one of the most common categories of water pollutants. Gasoline leakage from storage tank, transportation, pipelines and petrochemical wastewaters introduce these compounds into the environment, making surface waters and/or groundwaters unsuitable for many uses, including drinking [1]. BTEX (Benzene, toluene, ethylbenzene and xylene) are frequently detected in chemical and petrochemical wastewaters. These contaminants can cause adverse health effects to humans even at low concentrations [2]. Therefore their removal from groundwater and surface waters is a problem of great importance. Among several techniques developed for BTEX removal from waters, adsorption is one of the most efficient methods, thanks to satisfactory efficiencies even at low concentrations [3], easy operation and low cost [4]. Recently, high-silica zeolites have been shown to be environmental friendly materials able to efficiently sorb several organic pollutants from water, such as pharmaceuticals [5–7], polycyclic aromatic hydrocarbons [8], phenols [9] and petrol-derived compounds [10–12].

In literature, several works have focused on the advantages of zeolites as adsorbents, such as high selectivity and capacity, rapid kinetics, reduced interference from salt and humic substances, excellent resistance to chemical, biological, mechanical and thermal stress [9,13,14]. Even if zeolites are more expensive with respect to other adsorbents, they offer the possibility to be regenerated without loss of performances at relatively low temperatures, as demonstrated in previous works [10,15].

The investigation of several synthetic zeolites such as ZSM-5 [10], mordenite [4], ferrierite [16] and Y [11] for the removal of petrol-derived compounds from aqueous solutions showed that they are a promising material for water clean-up procedures. Another adsorbent that could be employed in such treatment is zeolite beta due to its large porosity and high surface area. Zeolite beta, indeed, has a three-dimensional intersecting channels system, two mutually perpendicular straight channels each with a cross section of 6.6 Å × 6.7 Å and a sinusoidal channel with a cross section of 5.6 Å × 5.6 Å [17]. This tortuous channels system is constituted by the intersection of the two main channels. The channel intersections of zeolite beta generate cavities whose sizes are in the order of 12–13 Å [17]. Crystallographic faults are frequently observed in beta zeolite and a structural model was proposed by Jansen et al. [18] to explain the creation of local defects by the connection of distorted layers. The structure of zeolite beta is disordered along [001] and it is related to three ordered structures by $a/3$ and/or $b/3$ displacements. The three ordered polytypes are designated frameworks A, B, and C [19,20]. Polytype A is tetragonal (space group P$4_1$22 or P$4_3$22, cell parameters $a = b \approx 12.5$ Å and $c \approx 26.4$ Å), polytype B is monoclinic (space group C2/c, cell parameters $a \approx b \approx 17.6$ Å, $c \approx 14.4$ Å and $\beta \approx 114°$), as well as Polytype C (space group P2/c, cell parameters $a \approx b \approx 12.5$ Å, $c \approx 27.6$ Å, and $\beta \approx 107°$).

It has been reported that thermal and hydrothermal treatments induce chemical and structural modifications in beta zeolites, for instance Trombetta et al. [21] observed that thermal treatments can cause dealumination and formation of extraframework aluminium species. The ease of dealumination of beta may be due to the presence of defect sites close to the framework aluminium which promotes bond hydrolysis, nonetheless the microporous structure is not affected by the loss of aluminium [22]. Other zeolites, such as ZSM-5 or mordenite, do not show significant crystallinity loss or dealumination after thermal treatment [10,15,22]. The precise structural modifications of beta zeolite are still a matter of research and the global effect of calcination on beta acidity is not totally clear, because of the presence of several types of acidic sites, with different acidity degree [23,24]. However, it can be inferred that beta zeolites could undergo to greater variations in adsorption properties due to calcination with respect to other zeolites. Also the hydrophilic/hydrophobic features, controlled by varying the SiO_2/Al_2O_3 ratio (SAR), can influence the behaviour of zeolites towards polar/non-polar reactants and products in adsorption and catalytic processes. In fact, the roles played both by calcination and by SAR on catalytic activity of beta zeolites received great attention [25,26]. However there are only few works dealing with the effects of both chemical the composition and thermal treatments of beta on the adsorption properties toward solutes from water solutions. Indeed, the phenomena observed in catalytic gas phase systems could be different from those in aqueous matrix, since it has already been reported that the presence of water can strongly interfere with organic compounds adsorption [27].

Therefore, the objective of this work is to investigate the adsorptive properties of beta zeolites (notated BEAs, since they belong to BEA framework type), with different Silica/Alumina ratios (SAR) before and after calcination for the removal of toluene (TOL) from aqueous solutions. The selected adsorbents were commercial beta zeolites: the possibility to find them on the market and to use them as-received from the manufacturer was considered a strong decision point for their selection.

2. Materials and Methods

2.1. Chemicals

Toluene (99% purity) was obtained from Sigma-Aldrich (Steinheim, Germany). High-performance liquid chromatography (HPLC) grade acetonitrile (ACN) was purchased from Merck (Darmstadt, Germany). The water was Milli-Q grade (Millipore, Billerica, MA, USA). Zeolite beta powders were obtained from Zeolyst International (Conshohocken, PA, USA) and their main characteristics are reported in Table 1.

Table 1. Zeolites characteristics.

Name	Product Code	SiO_2/Al_2O_3	Nominal Cation	Surface Area ($m^2 \cdot g^{-1}$)
Beta25	CP814E	25	Ammonium	680
Beta38	CP814C	38	Ammonium	710
Beta360	CP811C-300	360	Hydrogen	620

All the adsorbents were employed as-received (named Beta25, Beta38 and Beta360) and after a calcination process (referred to as Beta25c, Beta38c and Beta360c). Calcination was carried out by raising the temperature from room temperature to 600 °C in 1 h, then holding at 600 °C for 4 h. Finally, adsorbents were kept at room temperature for 3 h. Dry air circulation was maintained during both heating and cooling down. The calcined samples were kept in a desiccator and used within 2 days after thermal treatment.

2.2. Experimental

The adsorption isotherm was determined using the batch method. Batch experiments were carried out in duplicate in 20 mL crimp top reaction glass flasks sealed with polytetrafluoroethylene (PTFE) septa (Supelco, Bellefonte, PA, USA). The flasks were filled in order to have the minimum headspace and a solid/solution ratio of 1:2 ($mg \cdot mL^{-1}$) was employed. After equilibration, for 24 h at a temperature of 25.3 ± 0.5 °C under stirring, the solids were separated from the aqueous solution by filtration trough 0.22 µm polyvinylidene fluoride (PVDF) membrane filters purchased from Agilent Technologies (Santa Clara, CA, USA). The concentration of TOL was determined in the solutions before and after equilibration with zeolite by High Performance Liquid Chromatography/Diode Array Detection (HPLC/DAD) purchased from Waters (Waters Corporation, Milford, MA, USA).

2.3. Instrumentation

The HPLC/DAD was employed under isocratic elution conditions. The column (Agilent Technologies) was 150 mm × 4.6 mm, packed with a C18 silica-based stationary phase with a particle diameter of 5 µm and thermostated at 25 °C. The injection volume was 20 µL for all standards and samples. The mobile phase was a mixture ACN:H_2O 70:30 and the flow rate was 1 mL/min. Detection wavelength was set at 215 nm Thermogravimetric (TG), differential thermogravimetric (DTG) and differential thermal analyses (DTA) measurements of exhausted samples were performed in air up to 900 °C, at 10 °C·min^{-1} heating rate, using a simultaneous thermal analysis (STA) 409 PC LUXX®—NETZSCH Gerätebau GmbH (Verona, Italy). X-ray powder diffraction (XRPD) patterns of zeolites after TOL adsorption were measured on a Bruker (Billerica, MA, USA) D8 Advance Diffractometer equipped with a Si (Li)SOL-X solid-state detector. Statistical elaborations were carried out through MATLAB® ver. 9.1 software (The MathWorks Inc., Natick, MA, USA).

3. Results and Discussion

3.1. Adsorption from Aqueous Solutions

The adsorption kinetics was studied in order to obtain some important parameters, such as the kinetic constant, which allow the estimation of the time requested for reaching the equilibrium. Moreover, from kinetics measurements, qualitative information about the steps governing the adsorption process can be gained. The uptake q ($mg \cdot g^{-1}$) was calculated as follows:

$$q = \frac{(C_0 - C_e)V}{m} \tag{1}$$

where C_0 is the initial concentrations in solution ($mg \cdot L^{-1}$), C_e is the concentration at time t in kinetics experiments or at equilibrium ($mg \cdot L^{-1}$) for isotherm modelling, V is the solution volume (L) and m is the mass of adsorbent (g).

The kinetics was very fast for all the studied materials and the time to reach equilibrium was about 10 min. As an example, the uptake data obtained for TOL on Beta360 are shown in Figure 1. The data of Figure 1 were fitted by the pseudo-second order model (Equation (2)), which has been employed in many studies concerning the adsorption of organic compounds onto zeolites [28,29].

$$q_t = \frac{k_2 q^2 t}{1 + k_2 q t} \tag{2}$$

where q_t and q are the amounts of solute sorbed per mass of adsorbent at time t and at equilibrium, respectively, and k_2 is the second-order adsorption rate constant. The equilibrium uptake q and the adsorption rate constant k_2 were obtained from non-linear fit of q_t vs. t. Values of 3.39 (3.28, 3.50) and 0.46 (0.31, 0.62) were obtained for q_e and k_2, respectively: the confidence limits at 95% of probability are reported in brackets. The pseudo-second-order model fitted well all the sorption data as demonstrated by the resulting high coefficients of determination ($R^2 = 0.9915$). From Figure 1 it can also be seen that the surface adsorption (first part of the curve) is a faster process than the intraparticle diffusion of TOL into the zeolite micropores as alredy observed for ZSM-5 [10].

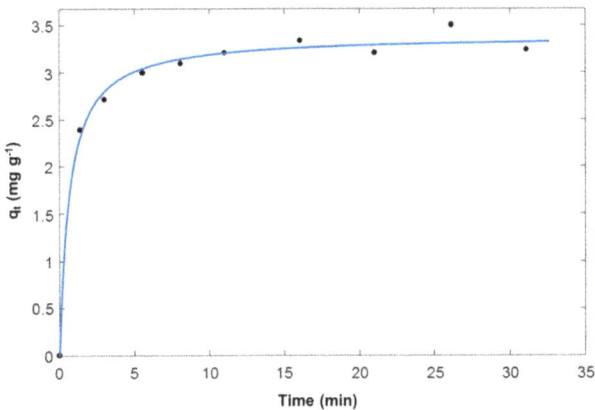

Figure 1. Adsorption kinetics of toluene (TOL) on Beta360: TOL uptake vs. contact time.

The relationship between the solute amount adsorbed for per unit mass of adsorbent q and its concentration at equilibrium C_e is provided by equilibrium adsorption isotherms. The Langmuir isotherm has been frequently used to describe the adsorption of organics in aqueous solutions onto hydrophobic zeolites [10,15,29]. This model considers a monolayer adsorption onto energetically equivalent adsorption sites and negligible sorbate–sorbate interactions. It can be represented by the following equation [30].

$$q = \frac{q_s b C_e}{1 + b C_e} \tag{3}$$

where b is the binding constant (L·mg^{-1}) and q_s is the saturation capacity of the adsorbent material (mg·g^{-1}). This model has already been employed for adsorption on BEAs of several classes of organic compounds, such as pharmaceuticals [6], etheramine [29], xylene isomers and ethylbenzene [31].

Freundlich isotherm is a relationship describing non-ideal and reversible adsorption, not restricted to the formation of monolayer. In fact, this empirical model can be applied to multilayer adsorption, with non-uniform distribution of adsorption heat and affinities over the heterogeneous surface [32]. The Freundlich isotherm model can be expressed as [33].

$$q = K_F C_e^{1/n} \tag{4}$$

where K_F is a constant indicative of the adsorption capacity and $1/n$ is a measure of the surface heterogeneity, ranging between 0 and 1. The surface heterogeneity increases as $1/n$ gets closer to zero. The Freundlich isotherm equation was found to have a better fit than the Langmuir equation for TOL adsorption on as-received BEAs (vide infra). This model was also used also by Wang et al. [28] for describing the adsorption of 1,3-propanediol on BEAs zeolites.

Another isotherm model employed to describe multiple adsorbate/adsorbent interactions is that proposed by Tóth [34].

$$q = \frac{q_S b C_e}{\left[1 + (bC_e)^v\right]^{1/v}} \tag{5}$$

where v is a parameter accounting for the heterogeneity of adsorption energies. If $v = 1$, the Tóth model corresponds to the Langmuir model [34].

The adsorption isotherms of TOL on both as-received and calcined BEAs are shown in Figure 2, where it can be noted that the isotherms shape of as-received and calcined beta zeolites are different from each other mainly due to modification on the adsorbate/surface interaction energy caused by calcination of the adsorbent. In particular, the thermal treatment considerably increases the adsorption efficiency of all BEAs toward TOL in the low concentration range. This finding has also been observed also for polar compounds such as pharmaceuticals [6]. It has been suggested that part of the adsorption properties of BEA zeolites originates from faults in the zeolitic structure [35]. In addition to its Brønsted acidity, beta zeolite also displays also Lewis acidity [36]. The calcination leads to the conversion of NH_4-BEA to H-BEA for Beta25 and Beta38 (see Table 1), as well as to structural and surface modifications for all the three beta zeolites [36,37]. In particular, the thermal treatment can lead to silanols condensation and, consequently, to the degradation of Brønsted acid sites by dehydroxylation. Together with the removal of water, the formation of Lewis acid sites occurs [24], as proposed by some studies [21,23] which found an increase in the ratio Lewis/Brønsted acid sites in the calcined material with respect to the as-received one. However, the global effect of calcination on beta acidity is not totally clear, because of the presence of several types of acidic sites, with different acidity degree [24,38]. It can be inferred that beta Lewis acid sites, whose formation has been promoted by thermal treatment, could interact with toluene as reported by Maretto et al. [39]. Therefore, calcination can lead to structural and compositional changes in beta zeolites, inducing to differences in adsorption properties [22,40]. The experimental data were fitted with all the three models (see Equations (3)–(5)). In order to compare these models, the statistical analysis of the fitting based of the square sum of errors and the number of parameters was performed. The isotherm parameters of the best fitted model estimated by non-linear fitting of the as-received and calcined BEAs are shown in Tables 2 and 3, respectively.

Table 2. Isotherm parameters for the adsorption of TOL on as-received BEAs estimated by non-linear fitting, according to the Freundlich model. The confidence limits at 95% of probability of the estimated parameters are reported in brackets.

As-Received Materials	K_F (mg·g^{-1})·(L·g^{-1})^n	n	R^2
Beta25	5.2 (3.5, 7.0)	0.86 (0.77, 0.95)	0.9953
Beta38	4.2 (2.5, 5.9)	0.79 (0.69, 0.89)	0.9936
Beta360	4.1 (2.8, 5.3)	0.93 (0.72, 1.1)	0.9956

Table 3. Isotherm parameters for the adsorption of TOL on calcined BEAs estimated by non-linear fitting, according to the Tóth model. The confidence limits at 95% of probability of the estimated parameters are reported in brackets.

Calcined Materials	q_s (mg·g^{-1})	b (L·mg^{-1})	v	R^2
Beta25c	234 (193, 275)	0.073 (0.043, 0.10)	0.96 (0.70, 1.2)	0.9584
Beta38c	224 (198, 250)	0.10 (0.075, 0.13)	0.91 (0.72, 1.1)	0.9688
Beta360c	241 (201, 280)	0.55 (0.30, 0.80)	0.84 (0.62, 1.0)	0.9667

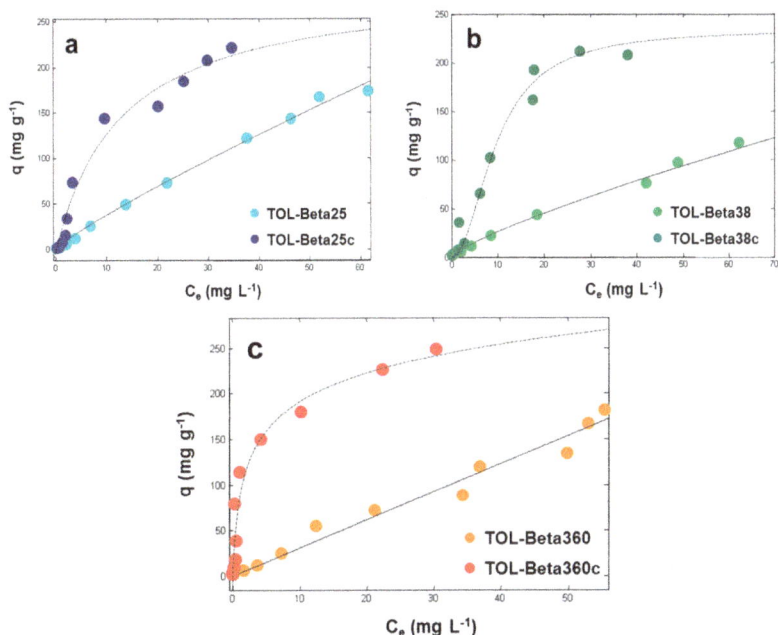

Figure 2. Adsorption isotherms of TOL on (**a**) Beta25 (light blue circles: as-received, dark blue circles: calcined); (**b**) Beta38 (light green circles: as-received, dark green circles: calcined) and (**c**) Beta360 (orange circles: as-received, red circles: calcined).

From Tables 2 and 3, it can be seen that as-received zeolites are fitted well by a Freundlich model, whereas the calcined materials are modelized by a Tóth isotherm equation. In particular, Table 2 shows that n constant for Beta360 is not statistically different, at 95% of probability, from 1, hence TOL adsorption on this zeolite follows a linear trend. On the contrary, values of n below 1 have been observed for both Beta25 and Beta38, indicating that the adsorbent surface is heterogeneous. The values of K_F found for the three as-received BEAs are not statistically different from each other at a probability of 95%. This finding may indicate similar adsorbent/adsorbate interactions, possibly due to the effect of physisorbed water on the zeolites porosities (see Section 3.2) and to the presence of structural defects in beta zeolites that make it difficult to assess the properties of the adsorption sites. By comparing calcined BEAs (Table 3), it can be seen that their saturation capacities are not statistically different at 95% of probability. High values of q_s were obtained for all the calcined adsorbents (above 20% w/w). This last finding makes calcined BEAs very promising as adsorbents in the remediation of contaminated waters at high concentration levels. Similar values of q_s were found in the adsorption of different organic contaminants on hydrophobic Y zeolite (FAU-type framework topology) [11,12]. However, the binding constants b obtained with Y zeolite were quite low, thus indicating that in the low concentrations range Y zeolite is generally less efficient than calcined BEAs. At low TOL concentrations, it has been proved that another hydrophobic zeolite, namely ZSM-5 (MFI-type framework topology) is very efficient [10]. In this case, in fact, the adsorption isotherm of TOL on ZSM-5 was characterized by a high binding constant (b was 3.17 ± 0.41), despite the lower saturation capacity of ZSM-5 than BEAs and Y (around 8% w/w). In the light of the above findings, it can be stated that calcined BEAs represent a good compromise for that which concerning TOL adsorption from aqueous solutions in a wide concentrations range. Concerning the binding constant b, Beta360c showed a higher value than those of Beta25c and Beta38c, which are not significantly different from each other at 95% of probability. This finding could be explained by considering that adsorption onto zeolites is driven by both electrostatic and non-covalent interactions [41]. It can be supposed that electrostatic interactions have a negligible contribution to the adsorption of an organophilic solute such as TOL, characterized by logK$_{ow}$ of 2.73. Therefore, it can be considered that the adsorption mechanism of TOL onto BEAs is driven mainly by non-covalent interactions, which become more relevant as SAR value increases.

3.2. Thermal and Structural Analyses

Thermogravimetric analysis were carried out for the as-received materials (i.e., Beta25, Beta38 and Beta360). A total weight loss of about 17% was observed for all the three samples for temperature up to 900 °C.

These weight losses can be divided up into two contributions: the first one at low temperature (i.e., lower than about 100 °C) due to the loss of water molecules weakly bonded to the zeolite surface and the second one at higher temperature mainly ascribable to the loss of ammonia from Beta25 and Beta38 as well as losses of structural water molecules and silanols condendation in all the beta samples (Figure 3).

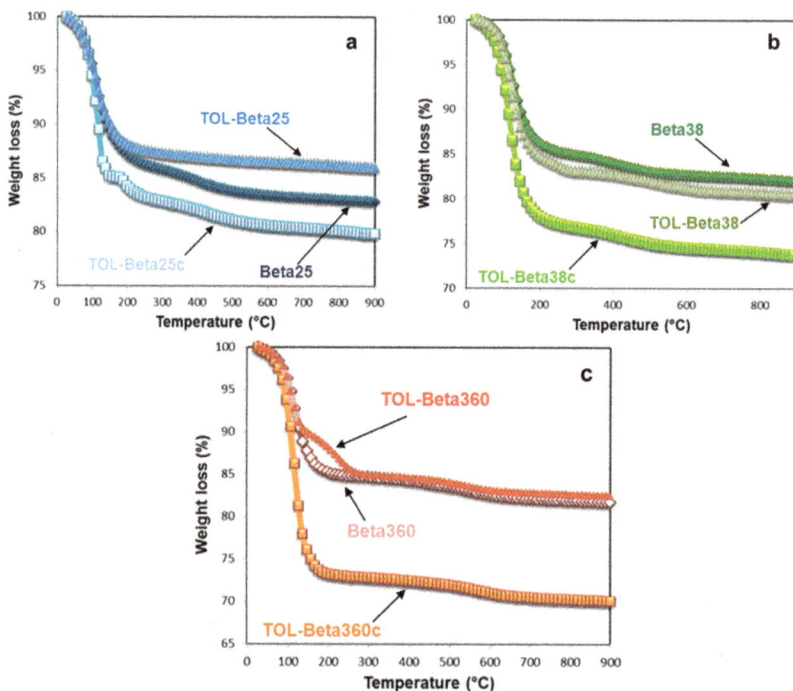

Figure 3. Thermogravimetric curves of as-received and calcined BEAs, before and after saturation with TOL: (**a**) Beta25; (**b**) Beta38; and (**c**) Beta360.

The TG analyses of calcined BEAs after TOL saturation show weight losses at 900 °C of 20.2%, 26.2% and 30% for Beta25c, Beta38c and Beta360c, respectively. However, these weight losses cannot be easily related to the adsorbed TOL amount since, as reported in Pasti et al. [6], the calcined zeolites can undergo to rehydratation process and the temperatures at which the adsorbed water and TOL are removed from the framework are very close to each other's. This makes it difficult to ascribe the whole weight loss to water or TOL alone. However, these results are in good agreement with the saturation capacities of the materials determined by adsorpion experiments (see Table 3).The X-ray powder diffraction patterns of both as-received and calcined Beta25, Beta38 and Beta360, before and after saturation with TOL are reported in Figure 4. By comparing the X-ray powder diffraction pattern of both the as-received and the calcined materials before and after TOL adsorption (see Figure 4) it can be observed that the peaks intensities in the low 2θ region change thus confirming the incorporation of molecules in the framework due to adsorption, moreover the differences in the patterns in the intermediate and high 2θ region indicates that the process is associated with the framework flexibility (expansion or contraction of the cell volume) [42,43]. Similar behaviour is also shown when the three zeolite samples before and after thermal treatment are compared.

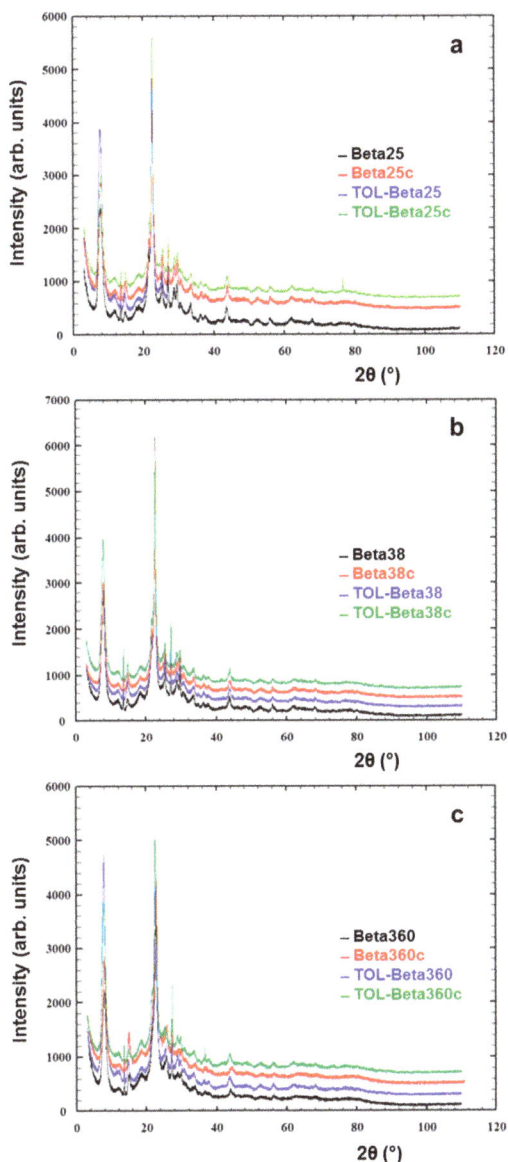

Figure 4. X-ray powder diffraction patterns of as-received and calcined BEAs before and after saturation with TOL: (**a**) Beta25; (**b**) Beta38; and (**c**) Beta360.

4. Conclusions

This work highlighted the differences in adsorption properties between as-received and calcined beta zeolites, with three different SARs, toward a water contaminant of great concern such as toluene. It has been observed that the calcination significantly improves the adsorption properties of all of the three zeolites.

The adsorption of toluene by calcined BEAs is characterized by high values of saturation capacity. The most hydrophobic calcined beta, i.e., Beta360c, showed the highest binding constant, thus indicating stronger adsorbent/adsorbate interactions than those of Beta25c and Beta38c. Consequently, Beta360 after thermal treatment is a promising adsorbent for the removal of toluene in water-containing systems. These results open new alternatives for the industrial application of this material, mainly in hydrocarbons adsorption processes in the presence of water.

Acknowledgments: Federico Giacometti is acknowledged for performing some of the experiments.

Author Contributions: Elena Sarti wrote the paper and performed some of the adsorption experiments with Tatiana Chenet; Luisa Pasti, Alberto Cavazzini and Annalisa Martucci conceived and designed the experiments and analyzed the data; Elisa Rodeghero performed thermogravimetric and X-ray data analysis.

Conflicts of Interest: The authors declare no conflicts of interest.

References

1. Aivalioti, M.; Vamvasakis, I.; Gidarakos, E. BTEX and MTBE adsorption onto raw and thermally modified diatomite. *J. Hazard. Mater.* **2010**, *178*, 136–143. [CrossRef] [PubMed]
2. Qu, F.; Zhu, L.; Yang, K. Adsorption behaviors of volatile organic compounds (VOCs) on porous clay heterostructures (PCH). *J. Hazard. Mater.* **2009**, *170*, 7–12. [CrossRef] [PubMed]
3. Gupta, V.K.; Verma, N. Removal of volatile organic compounds by cryogenic condensation followed by adsorption. *Chem. Eng. Sci.* **2002**, *57*, 2679–2696. [CrossRef]
4. Arletti, R.; Martucci, A.; Alberti, A.; Pasti, L.; Nassi, M.; Bagatin, R. Location of MTBE and toluene in the channel system of the zeolite mordenite: Adsorption and host–guest interactions. *J. Solid State Chem.* **2012**, *194*, 135–142. [CrossRef]
5. Martucci, A.; Pasti, L.; Marchetti, N.; Cavazzini, A.; Dondi, F.; Alberti, A. Adsorption of pharmaceuticals from aqueous solutions on synthetic zeolites. *Microporous Mesoporous Mater.* **2012**, *148*, 174–183. [CrossRef]
6. Pasti, L.; Sarti, E.; Cavazzini, A.; Marchetti, N.; Dondi, F.; Martucci, A. Factors affecting drug adsorption on beta zeolites. *J. Sep. Sci.* **2013**, *36*, 1604–1611. [CrossRef] [PubMed]
7. Braschi, I.; Blasioli, S.; Gigli, L.; Gessa, C.E.; Alberti, A.; Martucci, A. Removal of sulfonamide antibiotics from water: Evidence of adsorption into an organophilic zeolite Y by its structural modifications. *J. Hazard. Mater.* **2010**, *17*, 218–225. [CrossRef] [PubMed]
8. Costa, A.A.; Wilson, W.B.; Wang, H.; Campiglia, A.D.; Dias, J.A.; Dias, S.C.L. Comparison of BEA, USY and ZSM-5 for the quantitative extraction of polycyclic aromatic hydrocarbons from water samples. *Microporous Mesoporous Mater.* **2012**, *149*, 186–192. [CrossRef]
9. Khalid, M.; Joly, G.; Renaud, A.; Magnoux, P. Removal of phenol from water by adsorption using zeolites. *Ind. Eng. Chem. Res.* **2004**, *43*, 5275–5280. [CrossRef]
10. Rodeghero, E.; Martucci, A.; Cruciani, G.; Bagatin, R.; Sarti, E.; Bosi, V.; Pasti, L. Kinetics and dynamic behaviour of toluene desorption from ZSM-5 using in situ high-temperature synchrotron powder X-ray diffractionand chromatographic techniques. *Catal. Today* **2016**, *227*, 118–125. [CrossRef]
11. Martucci, A.; Braschi, I.; Bisio, C.; Sarti, E.; Rodeghero, E.; Bagatin, R.; Pasti, L. Influence of water on the retention of methyl tertiary-butyl ether by high silica ZSM-5 and Y zeolites: A multidisciplinary study on the adsorption from liquid and gas phase. *RSC Adv.* **2015**, *5*, 86997–87006. [CrossRef]
12. Pasti, L.; Martucci, A.; Nassi, M.; Cavazzini, A.; Alberti, A.; Bagatin, R. The role of water in DCE adsorption from aqueous solutions onto hydrophobic zeolites. *Microporous Mesoporous Mater.* **2012**, *160*, 182–193. [CrossRef]
13. Weitkamp, J. Zeolites and catalysis. *Solid State Ion.* **2000**, *131*, 175–188. [CrossRef]

14. Aivalioti, M.; Pothoulaki, D.; Papoulias, P.; Gidarakos, E. Removal of BTEX, MTBE and TAME from aqueous solutions by adsorption onto raw and thermally treated lignite. *J. Hazard. Mater.* **2012**, *207*, 136–146. [CrossRef] [PubMed]

15. Martucci, A.; Rodeghero, E.; Pasti, L.; Bosi, V.; Cruciani, G. Adsorption of 1,2-dichloroethane on ZSM-5 and desorption dynamics by in situ synchrotron powder X-ray diffraction. *Microporous Mesoporous Mater.* **2015**, *215*, 175–182. [CrossRef]

16. Martucci, A.; Leardini, L.; Nassi, M.; Sarti, E.; Bagatin, R.; Pasti, L. Removal of emerging organic contaminants from aqueous systems: Adsorption and location of methyl-tertiary-butylether on synthetic ferrierite. *Mineral. Mag.* **2014**, *78*, 1161–1175. [CrossRef]

17. Baerlocher, C.; Meir, W.M.; Olson, O.H. *Atlas of Zeolite Framework Types*, 5th revised ed.; Elsevier Science: Amsterdam, The Netherlands, 2001.

18. Jansen, J.C.; Creyghton, E.J.; Njo, S.L.; van Koningsveld, H.; van Bekkum, H. On the remarkable behaviour of zeolite Beta in acid catalysis. *Catal. Today* **1997**, *38*, 205–212. [CrossRef]

19. Higgins, J.B.; LaPierre, R.B.; Schlenker, J.L.; Rohrman, A.C.; Wood, J.D.; Kerr, G.T.; Rohrbaugh, W.J. The framework topology of zeolite beta. *Zeolites* **1988**, *8*, 446–452. [CrossRef]

20. Newsam, J.M.; Treacy, M.M.J.; Koetsier, W.T.; De Gruyter, C.B. Structural characterization of zeolite beta. *Proc. R. Soc. Lond. Ser. A* **1988**, *420*, 375–405. [CrossRef]

21. Trombetta, M.; Busca, G.; Storaro, L.; Lenarda, M.; Casagrande, M.; Zambon, A. Surface acidity modifications induced by thermal treatments and acid leaching on microcrystalline H-BEA zeolite. A FTIR, XRD and MAS-NMR study. *Phys. Chem. Chem. Phys.* **2000**, *2*, 3529–3537. [CrossRef]

22. Beyer, H.K.; Nagy, J.B.; Karge, H.G.; Kiricsi, I. *Catalysis by Microporous Materials*; Elsevier: Amsterdam, The Netherlands, 1995.

23. Otomo, R.; Yokoi, T.; Kondo, J.N.; Tatsumi, T. Dealuminated Beta zeolite as effective bifunctional catalyst for direct transformation of glucose to 5-hydroxymethylfurfural. *Appl. Catal. A Gen.* **2014**, *470*, 318–326. [CrossRef]

24. Čejka, J.; van Bekkum, H.; Corma, A.; Schüth, F. *Introduction to Zeolite Science and Practice*, 3rd revised ed.; Elsevier: Amsterdam, The Netherlands, 2007.

25. Al-Khattaf, S.; Ali, S.A.; Aitani, A.M.; Žilková, N.; Kubička, D.; Čejka, J. Recent advances in reactions of alkylbenzenes over novel zeolites: The effects of zeolite structure and morphology. *Catal. Rev.* **2014**, *56*, 333–402. [CrossRef]

26. Dědeček, J.; Sobalík, Z.; Wichterlová, B. Siting and distribution of framework aluminium atoms in silicon-rich zeolites and impact on catalysis. *Catal. Rev.* **2012**, *54*, 135–223. [CrossRef]

27. Krishna, R.; van Baten, J.M. Hydrogen bonding effects in adsorption of water–alcohol mixtures in zeolites and the consequences for the characteristics of the Maxwell–Stefan diffusivities. *Langmuir* **2010**, *26*, 10854–10867. [CrossRef] [PubMed]

28. Wang, Z.; Wu, Z.; Tan, T. Sorption equilibrium, mechanism and thermodynamics studies of 1,3-propanediol on beta zeolite from an aqueous solution. *Bioresour. Technol.* **2013**, *145*, 37–42. [CrossRef] [PubMed]

29. Magriotis, Z.M.; Leal, P.V.B.; de Sales, P.F.; Papini, R.M.; Viana, P.R.M.; Augusto Arroyo, P. A comparative study for the removal of mining wastewater by kaolinite, activated carbon and beta zeolite. *Appl. Clay Sci.* **2014**, *91*, 55–62. [CrossRef]

30. Langmuir, I. The adsorption of gases on plane surfaces of glass, mica and platinum. *J. Am. Chem. Soc.* **1918**, *40*, 1361–1403. [CrossRef]

31. Dehkordi, A.M.; Khademi, M. Adsorption of xylene isomers on Na-BETA zeolite: Equilibrium in batch adsorber. *Microporous Mesoporous Mater.* **2013**, *172*, 136–140. [CrossRef]

32. Foo, K.Y.; Hameed, B.H. Insights into the modeling of adsorption isotherm systems. *Chem. Eng. J.* **2010**, *156*, 2–10. [CrossRef]

33. Freundlich, H.M.F. Over the adsorption in solution. *J. Phys. Chem.* **1906**, *57*, 385–471.

34. Tóth, J. State equations of the solid gas interface layer. *Acta Chem. Acad. Sci. Hung.* **1971**, *69*, 311–317.

35. Camblor, M.A.; Pérez-Pariente, J. Crystallization of zeolite beta: Effect of Na and K ions. *Zeolites* **1991**, *11*, 202–210. [CrossRef]

36. Kunkeler, P.J.; Zuurdeeg, B.J.; van der Waal, J.C.; van Bokhoven, J.A.; Koningsberger, D.C.; van Bekkum, H. Zeolite Beta: The relationship between calcination procedure, aluminum configuration, and lewis acidity. *J. Catal.* **1998**, *180*, 234–244. [CrossRef]

37. Lohse, U.; Altrichter, B.; Fricke, R.; Pilz, W.; Schreier, E.; Garkisch, C.; Jancke, K. Synthesis of zeolite beta Part 2—Formation of zeolite beta and titanium-beta via an intermediate layer structure. *J. Chem. Soc. Faraday Trans.* **1997**, *93*, 505–512. [CrossRef]

38. Kiricsi, I.; Flego, C.; Pazzuconi, G.; Parker, W.O.; Millini, R.; Perego, C.; Bellussi, G. Progress toward understanding zeolite β acidity: An IR and ^{27}Al NMR spectroscopic study. *J. Phys. Chem.* **1994**, *98*, 4627–4634. [CrossRef]

39. Maretto, M.; Vignola, R.; Williams, C.D.; Bagatin, R.; Latini, A.; Petrangeli Papini, M. Adsorption of hydrocarbons from industrial wastewater onto a silica mesoporous material: Structural and thermal study. *Microporous Mesoporous Mater.* **2015**, *203*, 139–150. [CrossRef]

40. Burke, N.R.; Trimm, D.L.; Howe, R.F. The effect of silica:alumina ratio and hydrothermal ageing on the adsorption characteristics of BEA zeolites for cold start emission control. *Appl. Catal. B Environ.* **2003**, *46*, 97–104. [CrossRef]

41. Krohn, J.E.; Tsapatsis, M. Amino acid adsorption on zeolite β. *Langmuir* **2005**, *21*, 8743–8750. [CrossRef] [PubMed]

42. Sánchez-Gil, V.; Noya, E.G.; Sanz, A.; Khatib, S.J.; Guil, J.M.; Lomba, E.; Marguta, R.; Valencia, S. Experimental and simulation studies of the stepped adsorption of toluene on pure-silica MEL zeolite. *J. Phys. Chem. C* **2016**, *120*, 8640–8652. [CrossRef]

43. Dong, J.; Tian, T.; Ren, L.; Zhang, Y.; Xu, J.; Cheng, X. CuO nanoparticles incorporated in hierarchical MFI zeolite as highly active electrocatalyst for non-enzymatic glucose sensing. *Colloid Surf. B* **2015**, *125*, 206–212. [CrossRef] [PubMed]

minerals

MDPI

Article

Use of Spent Zeolite Sorbents for the Preparation of Lightweight Aggregates Differing in Microstructure

Wojciech Franus [1,*], **Grzegorz Jozefaciuk** [2], **Lidia Bandura** [1] and **Małgorzata Franus** [1]

[1] Department of Geotechnical Science, Faculty of Civil Engineering and Architecture,
  Lublin University of Technology, Nadbystrzycka 40, 20-618 Lublin, Poland; l.bandura@pollub.pl (L.B.);
  m.franus@pollub.pl (M.F.)
[2] Department of Physical Chemistry of Porous Materials, Institute of Agrophysics,
  Polish Academy of Sciences, Doświadczalna 4, 20-290 Lublin, Poland; g.jozefaciuk@ipan.lublin.pl
* Correspondence: w.franus@pollub.pl; Tel.: +48-81-538-4416

Academic Editor: Peng Yuan
Received: 9 January 2017; Accepted: 12 February 2017; Published: 17 February 2017

Abstract: Lightweight aggregates (LWAs) made by sintering beidellitic clay deposits at high temperatures, with and without the addition of spent zeolitic sorbents (clinoptilolitic tuff and Na-P1 made from fly ash) containing diesel oil, were investigated. Mineral composition of the aggregates determined by X-ray diffraction was highly uniformized in respect of the initial composition of the substrates. The microstructure of the LWAs, which were studied with a combination of mercury porosimetry, microtomography, nitrogen adsorption/desorption isotherms and scanning electron microscopy, was markedly modified by the spent zeolites, which diminished bulk densities, increased porosities and pore radii. The addition of zeolites decreased water absorption and the compressive strength of the LWAs. The spent Na-P1 had a greater effect on the LWAs' structure than the clinoptilolite.

Keywords: lightweight aggregate; spent sorbents; petroleum; mercury porosimetry; microtomography; porosity

1. Introduction

Lightweight aggregates (LWAs) are building materials, produced from different minerals (including ordinary soil clay, perlite, vermiculite, and natural and synthetic zeolites) by rapid sintering/heating at high temperatures up to 1300 °C [1]. To achieve expanded material appropriately, two conditions are necessary: the presence of substances that release gases at high temperature, and a plastic phase with adequate viscosity, which is able to trap the released gases [2]. The expanded clay aggregates are non-flammable and highly resistant to chemical, biological and weather conditions. Their highly porous structure is represented mainly by closed pores surrounded by glassy coatings, which are formed during the thermal transformation of clay minerals. As a consequence, LWAs have relatively low particle and bulk densities, low thermal conductivity and sound dampening characteristics [3–8], thereby allowing them to have broad applications in the construction industry, geotechnics, gardening and agriculture [4,5,9–16].

Much effort has been recently invested to reuse different kinds of waste materials, in order to avoid their disposal in landfills and paying additional environmental taxes, as well as to reduce production costs [17,18]. Many waste materials, such as combustion ashes [19], waste glass [15], sewage or industrial sludge [20–23], incinerator bottom ash [24], mining residues, heavy metal sludge, washing aggregate sludge [4], polishing residue, lignite coal fly ash [25,26], spent adsorbents [27,28] and contaminated mine soil [29], have been used as additives for the production of LWAs. Some of these materials can contribute to the foaming or bloating that occurs during LWAs' sintering, thus increasing their porosity.

Among their wide industrial and environmental applications [30–33], zeolite minerals have recently been described as very efficient sorbents, especially for the cleanup of oil land spills [34–39]; however, significant amounts of waste materials are produced in parallel. According to our knowledge, waste zeolitic sorbents containing petroleum products have not yet been considered for LWA production. As evidenced by several research studies, the mineral composition and organic amendments to the substrate can determine the physical properties of LWAs. Therefore, we hypothesized that the addition of waste zeolites will modify the structure of the standard clay-based LWAs towards higher porosity, which will differ depending on the zeolite used.

2. Materials and Methods

2.1. Substrates

The starting materials for LWA preparation were beidellitic clay deposit (Budy Mszczonowskie, Poland) and two spent zeolitic sorbents: a natural clinoptilolitic tuff (Socirnica, Ukraine) [40] and the synthetic Na-P1 obtained by hydrothermal conversion of fly ash with sodium hydroxide, according to Wdowin et al. [41]. Both zeolites were dried at 105 °C, milled in a rotary mill to < 0.1 mm diameter and enriched to their maximum sorption capacities (25% w/w for the tuff and 50% w/w for Na-P1) with the diesel fuel Verva ON taken from the Orlen petrol station. The sorption capacities were measured by soaking the zeolites in the fuel drop by drop and weighing.

2.2. Lightweight Aggregates Preparation

90 g (90% w/w) of the clay samples were admixed with 10 g (10% w/w) of the spent sorbents, carefully homogenized and wetted with water (around 40 mL/100 g of the dry mass) (drop by drop at the end) in order to obtain plastic masses at the plastic limit state, according to ASTM D 4318 [42]. From these masses, granules of around 15 mm were formed by hand, air-dried at room temperature for 24 h, then at 50 °C for 2 h and finally at 105 °C for 12 h. The dry granules were placed into the SM-2002 "Czylok" laboratory furnace, subjected to heating up to 1170 °C with 5 °C·min^{-1} temperature increase, sintered at 1170 °C for 30 min and left in the furnace overnight for cooling to approximately 100 °C. The cooled LWAs were stored in closed vessels. The aggregates prepared from the natural clay deposit will be abbreviated further as CLAY, with those admixed with the clinoptilolitic tuff as CLIN and those with Na-P1 as NAP1.

2.3. Methods of Characterization

All measurements described below were performed in triplicate and all data presented further are averages from these replicates.

2.3.1. Mineralogical and Physical Properties

Mineralogical composition of the substrates and the obtained LWAs was examined by X-ray diffraction analysis using a X'pert PROMPD spectrometer with a PW 3050/60 goniometer (Panalytical, Almelo, The Netherlands), Cu lamp and graphite monochromator within a 2θ range of 5°–65°. Identification of the mineral phases was based on the JCPDS-ICDD database.

Solid phase density (SPD) was measured by water pycnometry for finely crushed (<0.1 mm) aggregates.

The particle density (BD) of the aggregates was estimated from their volume (measured by immersion in mercury) and mass (weighing).

Water absorption WA_{24h} was determined after soaking LWAs for 24 h in water and weighing according to EN-ISO 1097-6 [43].

The compressive strength *Ca*, being the force necessary to pass a piston for a certain depth into a cylinder filled with the studied material, was determined according to UNE-EN 13055-1/AC:2004-10-22 [44].

The freezing resistance F of the aggregates, which express the percentage loss of the mass of the aggregate soaked in water and subjected to 10 cycles of freezing-thawing (−17.5 and 20 °C, respectively) was determined by the UNE-EN 12697-2:1999 standard [45].

The laser diffraction method was applied in order to measure the particle size distribution of the initial materials subjected to 300 W of ultrasonication for 1 min using a Mastersizer 2000 with a Hydro G dispersion unit provided by Malvern UK. When obscuration after adding the sample to the measuring system exceeded 10%–20%, it was lowered by using the procedure that ensures there is no discrimination of any fraction [46]. For the solid phase, the refraction index was taken as 1.52 and the absorption index as 0.1; for water, the refraction index was taken as 1.33.

2.3.2. Structural Characteristics

X-ray computational microtomography was applied for 3D scanning of the studied LWAs using a Nanotom S device (General Electrics, Frankfurt, Germany). The X-ray source with a molybdenum target, operated at a cathode current of 230 μA and a 60 kV voltage was used for X-ray generation. The scanning process consisted of two stages: an initial pre-scan and a main measurement scan. Prior to the final measurement scan, each sample was subjected to a short 40 min pre-scan in order to heat it up and reach thermal stability, which was maintained further during the main scan lasting 150 min. The scanned specimens were dry, so the only effect of heating by X-rays on the measurement could have been caused by the thermal elongation of the sample holder. The pre-scan eliminated this problem. During the main scan, 2400 2D cross-sectional images were acquired with a spatial resolution (voxel size) of about 0.0063 mm and then used for 3D porous space reconstruction. The resulting 3D 16 bit grey-level images represent the spatial structure of specimens. Image analysis techniques were used for further processing. Initially, the bit depth of images was reduced from 16 to 8 bit. After that, a 3D median filter including a uniform kernel with a diameter equal to 3 px was used for noise reduction. The next step was the thresholding procedure, which utilized the Otsu algorithm. Threshold images had a 1 bit color depth with black areas representing pores. These preprocessing steps were performed using ImageJ software. For further analysis, Avizo software was used. The 3D watershed-based segmentation algorithm and then the labelling algorithm were used to separate the connected pores into individual ones. Geometrical characteristics of the pores including equivalent diameter (a diameter of the sphere with the same volume as a pore), volume, surface and fractal dimension of pores according to the maximal ball method [47] were calculated from three 3D images.

Mercury intrusion porosimetry (MIP) tests were performed for pressures ranging from about 0.1 to 200 MPa (pore radii ranged from about 10.0 to 3.8×10^{-3} μm). The intrusion volumes were measured at stepwise increasing pressures, which allowed for equilibration at each pressure step. The maximum deviations between the mercury intrusion volumes were no higher than 6.9% and occurred mainly at low pressures (largest pores). The volume of mercury V [$m^3 \cdot kg^{-1}$] intruded at a given pressure P [Pa] gave the pore volume that can be accessed. The intrusion pressure was translated into an equivalent pore radius R [m] following the Washburn equation:

$$P = -A \cdot \sigma_m \cdot \cos\alpha_m / R \qquad (1)$$

where σ_m is the mercury surface tension (0.485 $N \cdot m^{-1}$), α_m is the mercury/solid contact angle (taken as 141.3° for all studied materials) and A is a shape factor (equal to 2 for the assumed capillary pores). The total range for the pore radii in the mercury intrusion curve was divided into sections in steps of 0.1 $\log(R)$.

Knowing the dependence of V vs. R, a normalized pore size distribution, $\chi(R)$, was calculated and expressed in the logarithmic scale [48]:

$$\chi(R) = 1/V_{max}\cdot dV/d\log(R). \tag{2}$$

By knowing $\chi(R)$, the average pore radius, R_{av}, was calculated from:

$$R_{av} = \int R\cdot\chi(R)\cdot dR \tag{3}$$

If a range of pore sizes, wherein the pore volume depends on a power of the pore radius, could be found, this was interpreted in terms of pore surface fractal behavior. In this case, the dependence of $\log(dV/dR)$ against $\log R$ was plotted and, from the slope of its linear part, the fractal dimension of ·pore surface D was derived according to Pachepsky et al. [49]:

$$Ds = 2 - slope \tag{4}$$

To define the linear range of fractality, the Yokoya et al. [50] procedure was applied. According to this procedure the measure of the linearity L for the set of the points in a x-y plane is:

$$L = (4\sigma^2_{xy} + (\sigma_{yy} - \sigma_{xx})^2)^{1/2}\cdot(\sigma_{yy} + \sigma_{xx})^{-1} \tag{5}$$

where σ_{xx}, σ_{yy} and σ_{xy} are the variances of x-coordinates, y-coordinates and the covariance between x and y coordinate sets, respectively.

The L value falls between 0 (for uncorrelated and random points) and 1 (for points on a straight line). To separate out the linearity range, the value of L is computed for the first three points, then for the first four, five and so on until the value of L increases. The end of the linearity range is within the points after which the value of L begins to decrease. From the estimated linearity range, the two first and/or two last points were rejected if this caused an increase in the linear regression coefficient between the considered data.

The apparent solid phase skeletal densities of the samples SSD (which are lower than true solid phase densities due to the residence of the finest pores in the solid phase that are not filled by mercury at its highest pressure) and the total surface of MIP available pores $S_{(MIP)}$ were calculated by the porosimetric data analysis program provided by the equipment manufacturer.

Nitrogen adsorption isotherms were measured at the temperature of liquid nitrogen using ASAP 2020MP manufactured by Micromeritics (Norcross, GA, USA).

The scanning electron microscope (SEM) images of the tested materials were taken using an FEI Quanta 250 FEG microscope equipped with the energy dispersion scattering EDS-EDAX system for chemical composition analysis(FEI, Hilsboro, OR, USA). From three SEM images, the sizes of the finest pores were estimated using the Aphelion 4.0.10 image analysis package and the Vogel and Roth procedure [51].

3. Results and Discussion

3.1. Mineralogical and Physical Properties

Figure 1 illustrates the particle size distribution of the initial materials.

The clay material is composed of the finest particles with an average diameter of 58 μm, followed by NaP1 with a similar average diameter of 52 μm. The largest particles occur in natural clinoptilolitic tuff, for which the average diameter is 178 μm.

The main mineral components of the raw clay material were around 51% of beidellite (d_{hkl} 15.15, 4.44, 2,59 and 1.49 Å), 24% of quartz, 9% of kaolinite (d_{hkl} 7.14, 4.48 and 4.36 Å), 7% of illite (d_{hkl} 10.01, 5.02, 4.48, 3.34, 2.59 and 1.49 Å), 7% of feldspars (d_{hkl} 3.19, 3.68 and 4.22 Å) and less than 2% of iron hydroxides (Figure 2). The main mineral component of the clinoptilolitic tuff was clinoptilolite as recognized by d_{hkl} = 8.95, 7.94, 3.96 and 3.90 Å XRD reflections. The presence of the Na-P1 phase in the product of fly ash conversion was confirmed by d_{hkl} = 7.10, 5.01, 4.10 and 3.18 Å.

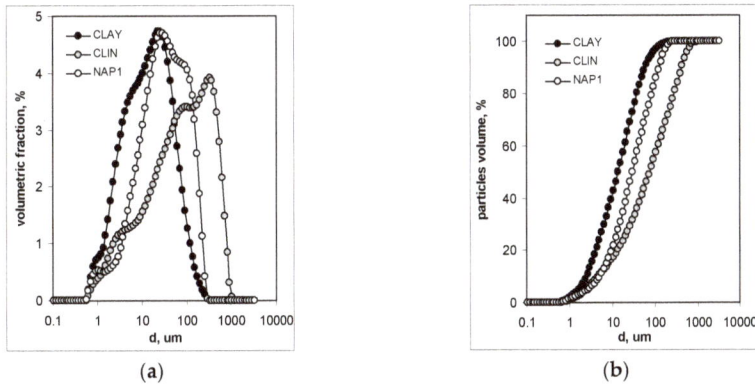

(a)   (b)

Figure 1. Particle size (diameter) distributions for the initial materials (**a**) and cumulative volume (scaled to 100%) versus particle diameter plot (**b**).

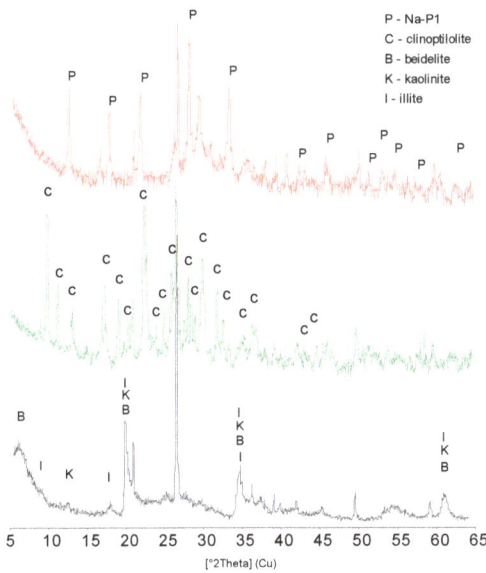

Figure 2. X-ray diffraction (XRD) patterns of the original clay deposit (**black**), clinoptilolitic tuff (**green**) and zeolite Na-P1 (**red**).

The XRD patterns of LWAs obtained from the clay and a mixture of clay and spent zeolitic sorbents are shown in Figure 3.

An extremely high degree of uniformization of the mineral composition of the sintered substrates is observed in the XRD spectra of the LWAs produced from different materials. The main mineral components of all LWAs are mullite (d_{hkl} 3.39, 5.41, 3.42 and 2.21 Å) and quartz (3.34, 4.25 and 1.81 Å). The presence of mullite is an effect of the melting of the original clay minerals (beidellite, illite, kaolinite) [52]. One can observe that iron hydroxides were transformed into well-defined hematite (d_{hkl} 2.70 and 2.51 Å), while the feldspars remained intact. Apart from the defined mineral phases, a significant contribution of an amorphous glassy phase can be distinguished by the rise in the background line within the range 15°–30° 2θ, which was the highest for LWAs admixed with Na-P1.

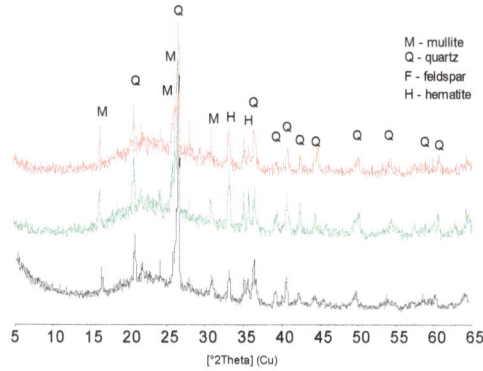

Figure 3. XRD patterns of the lightweight aggregates (LWAs): CLAY (**black**), CLIN (**green**) and NAP1 (**red**).

The physical properties of the studied LWAs are shown in Table 1. The solid phase density decreases slightly due to the spent sorbents addition, whereas the particle density decreases markedly, indicating the effect of the spent zeolites on the aggregates' expansion. Despite the decrease in particle density suggesting much larger porosity, the water absorption decreases for all LWAs produced with spent sorbents admixtures. Taking into account the similar mineral composition of the LWAs, resulting most probably in similar surface properties (wettability/hydrophobicity) of the aggregates material, the above differences may be connected to differences in the pore system, particularly in the amount of closed pores unavailable for water. According to Hung and Hwang [53], a particle with isolated pores or a vitrified surface tends to absorb less water than one having connected or open pores. Water absorption of the studied LWAs containing the spent sorbents is markedly lower than for several commercial ones as stated by their manufacturers, such as Lytag (17.55%), Arlita (20%) and Leca (30.3%).

Table 1. Physical parameters of the LWAs.

Parameter	CLAY	CLIN	NAP1
Solid phase density SPD, g·cm^{-3}	2.71	2.63	2.59
Particle density BD, g·cm^{-3}	1.74	1.27	0.76
Water absorption WA_{24}, %	20	10.00	11.5
Frost resistance F, %	<1	<1	<1
Compressive strength C_a, MPa	3.4	1.56	1.41

Lower water absorption may have technological advantages for building purposes. The frost resistance test showed that all LWAs lost not more than 1% mass after freezing that indicates their high resilience against variations in climate conditions. The aggregate grains did not show any occurrence of cracks after the test, probably because water penetrating the grains has not filled their whole pore space, meaning that it did not cause any visible aggregate damage after freezing. The compressive strength of the studied LWAs significantly decreased after the addition of spent sorbents. However, their mechanical resistance is still higher than that of some commercially available LWAs, such as Lytag (0.43 MPa), Leca (0.09 MPa) [54] or Leca Weber (0.75 MPa) (Saint-Gobain Construction Products

Poland) and markedly higher than 0.44 MPa, which is the internationally accepted standard for solid waste materials used for land levelling [55].

3.2. Structure Characteristics

Exemplary microtomography cross sections of the studied LWAs are presented in Figure 4, wherein quite different porous structures of the studied materials are seen.

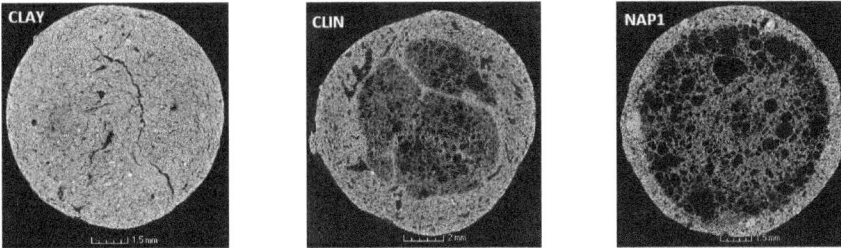

Figure 4. Exemplary 2D cross-sectional images derived from microtomography for the studied materials. Black areas: pores; white areas: solid.

On the external surfaces of both aggregates containing the diesel oil, a well-developed vitrified layer is seen. However, Gonzáles-Corrochano et al. [5] did not observe the formation of such a layer in LWAs manufactured with used motor oil. The visual analysis of the scans reveals that the LWAs have thick, dense areas, which extend throughout the whole CLAY aggregate, while being limited to the external layer for CLIN and NAP1, for which it is the thinnest. Most probably the thickness of this layer depends on the oil content. More oil evolves more gases during sintering and the resulted more porous structure reduces the number and increases the distance of connections between the molten solid thus its condensation is limited to smaller external space. It is also possible that more time is needed to decompose more oil what provides less time for solid condensation.

Calculated from 3D scans, the pore volume vs. pore radius dependencies and pore size distribution functions are presented in Figure 5.

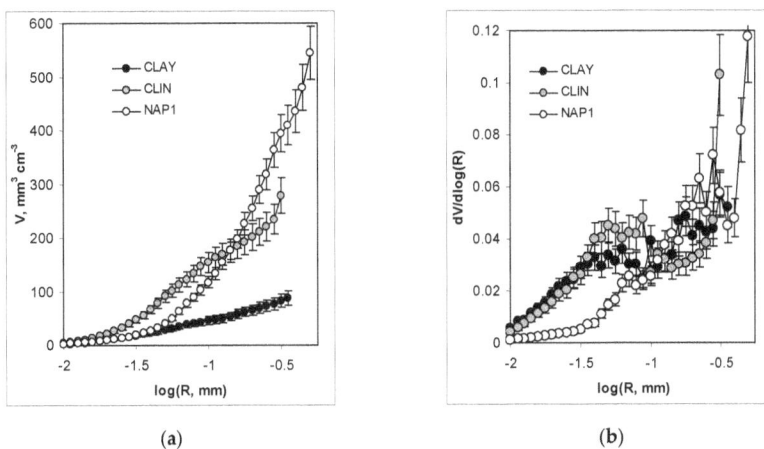

(a) (b)

Figure 5. Pore volume vs. pore radius dependencies (**a**) and normalized pore size distribution functions; (**b**) derived from microtomography scans. The points show average results, while the error bars show differences between the average and the experimental replicates.

As it is seen in Figure 5a,b, the LWA with used Na-P1 develops the largest pores and the largest pore volumes, particularly in the range of large pores. The volume of small pores is similar for NAP1 and CLAY, whereas CLIN possesses the largest volume of these pores. Pore size distribution functions (Figure 5b) show that CLIN aggregates contain the highest amount of pores lower than 0.1 mm, whereas the lowest amount of these pores is found in NAP1 aggregates.

MIP curves relating the intruded mercury (pore) volume to the logarithm of the pore radius and the normalized pore size distribution functions for the studied materials are presented in Figure 6. It is worth noting that the mercury extrusion branches (data not shown) were, in all cases, practically parallel to the log(R)-axis, indicating that practically all the mercury is accumulated in the pore voids and that the amount of the necks (channels) connecting these voids is negligible.

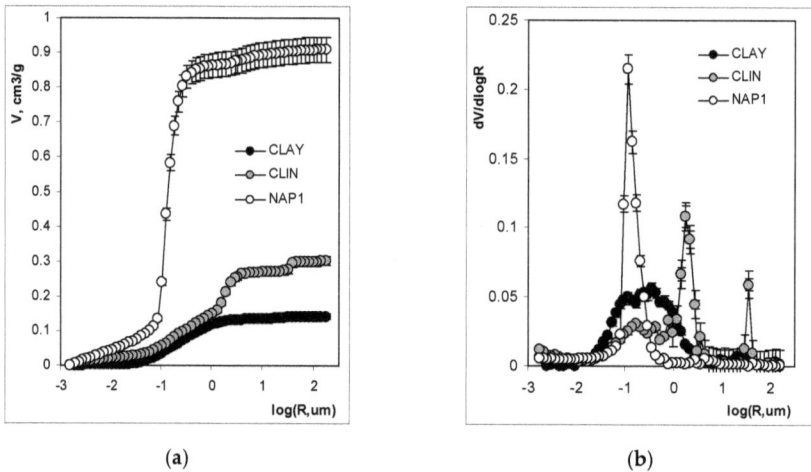

(a)   (b)

Figure 6. MIP curves (a) and normalized pore size distribution functions; (b) for the studied aggregates. The points show average results, while the error bars show maximum differences between the average and the experimental replicates.

The volume of intruded mercury (Figure 6a) is the lowest for the LWA containing only the original clay deposit, the intermediate for that enriched with the spent clinoptilolite, and the highest for the material containing spent NaP1. The pore size distributions for CLAY and NaP1 are unimodal (Figure 6b). One broad peak is noted for CLAY with the maximum located at around 0.32 μm (log$R$ ~ 0.5), while one narrow peak is found for NaP1 with maximum at R ~ 0.16 μm. Three peaks on the pore size distributions (PSD) function of CLIN are present: two narrow peaks at 32 and 2.5 μm, and one broad peak at around 0.16 μm. In contrast, Korat et al. [10] observed only bimodal MIP pore size distributions (peaks with maximum values between 0.1 and 1 μm, above 10 μm, and up to 100 μm) for LWAs prepared from fly ash obtained from coal combustion and silica sludge.

Comparing the pore size distribution functions derived from MIP and microtomography, one can see that MIP measurements allocate the sizes of almost the entire volume of the pores towards an underestimation of the large pores and an overestimation of the small pores. This phenomenon, as summarized by Korat et al. [10], appears to be rather intrinsic than accidental, which derives from the lack of direct accessibility for most of the pore volume (including air voids) to the mercury surrounding the specimen. Furthermore, in the case of highly porous structures, errors can also be made due to the breaking of the inner pore's walls, which then give distorted results.

Fractal plots for the studied materials are illustrated in Figure 7. As a rule, the fractal behaviour of the porosity of natural objects occurs in a limited range of pore dimensions (called upper and lower cut-offs) [49].

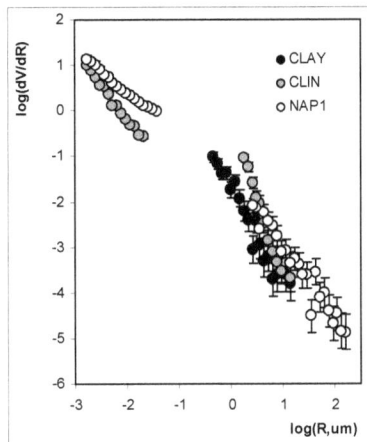

Figure 7. Fractal plots for the studied materials. The points show average results, while the error bars show maximum differences between the average and the experimental replicates.

The geometrical irregularities and roughness of the pore surface have an essential influence on the value of the fractal dimensions, which, for porous solids, may vary from 2 to 3. The lower limiting value of 2 corresponds to a perfectly regular pore surface, whereas the upper limiting value of 3 relates to the maximum allowed pore surface complexity [49]. The linearity ranges of log-log plots of dV/dR vs. R can be found for the studied aggregates. One linearity range is found for CLAY in the range of large pores. NAP1 and CLIN exhibit two ranges of linearity: one for large pores and the second for narrow pores.

However, the slopes of the linear log-log plots are very high in all cases, such that the calculated fractal dimensions of the pore surfaces are larger than 3 in all cases except for narrow pores of NAP1 (see Table 2 below). This may result from the specific structure of the aggregates. The large pore voids are accessible through markedly narrower entrances, therefore the volume of mercury forced into a large pore is attributed to the radius of the entrance and not to the radius of the void, falsifying the location of pore volume. In fractal dimension calculations a cylindrical pore model was applied assuming that the pore is a long capillary having the radius of the entrance. It is far from reality and leads to rapid increase of pore volume with pore (entrance) radius. Having such high increase in pore volume V vs. radius R dependence the cylindrical pore model calculates high dV/dR values that gives fractal dimension D values higher than 3.

Extremely low nitrogen adsorption and the calculated surface areas of the produced LWAs, which are less than 1 m^2/g (see Table 2), indicate that either the vitrified layer produced during heating has an extremely flat surface or the closed intra-aggregate pores are not available for nitrogen molecules.

SEM microphotographs of the obtained LWAs presented in Figure 8 show differences in the finest pores' structure of the aggregates. LWA prepared from clay is characterized by a compact texture, with the smallest pores being oval and frequently elongated. Aggregates with the admixtures of spent zeolites have pores of larger sizes, being the largest ones for NAP1. The placement of the pores is rather irregular.

Table 2. Pore parameters of the studied LWAs.

Pore Parameter	Unit	CLAY	CLIN	NAP1
Data from microtomography				
Total pore volume	$cm^3 \cdot g^{-1}$	0.044	0.223	0.719
Average pore radius	μm	130	120	250
Porosity (v/v)	%	8.7	27.9	54.5
Fractal dimension	/-	2.66	2.74	2.74
Data from MIP				
Total pore volume	$cm^3 \cdot g^{-1}$	0.141	0.301	0.909
Total Pore Area	$m^2 \cdot g^{-1}$	3.9	16.7	41.7
Average pore radius	μm	2.37	5.27	0.99
Particle density	$g \cdot cm^{-3}$	1.77	1.30	0.74
Solid skeletal density SSD	$g \cdot cm^{-3}$	2.35	2.14	2.23
Porosity (v/v)	%	25.0	39.2	66.9
Fractal dimension for large pores	/-	3.56	3.51	3.61
Fractal dimension for narrow pores	/-	N/a	3.5	2.91
Data from SEM				
Pore area (volume)	%	17	23	31
Dominant pore radius	μm	3	5	40

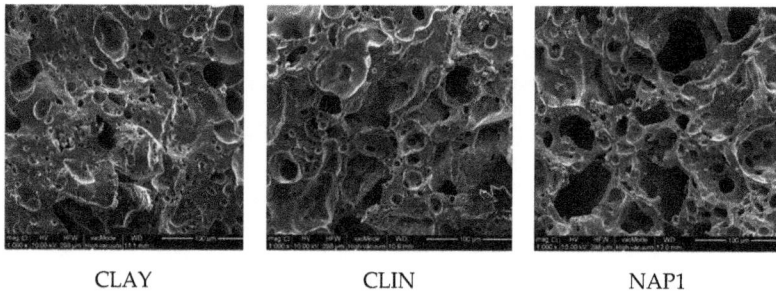

CLAY   CLIN   NAP1

Figure 8. Representative scanning electron microscope microphotographs (SEM) of the studied aggregate sections.

The pore parameters of the pore system derived from microtomography, MIP and SEM experiments are summarized in Table 2.

All methods applied give the highest pore volumes and porosities for NAP1 and the lowest for CLAY aggregates. As a rule, MIP measures significantly higher pore volumes and porosities than microtomography. The measuring range of microtomography starts from ~6 μm upwards, while it runs from ~4 nm to ~14 μm for MIP; at the first glance, it does not seem possible that MIP registers larger porosities. However, mercury can invade the whole aggregate interior through narrow entrances to the large pores, thereby filling all large pores inside. Therefore one can state that the total porosity values measured by microtomography are more reliable than these derived from MIP. LWAs made from the clay deposit have the smallest porosity and the highest particle density, whereas LWAs containing spent clinoptilolite and NaP-1 zeolites have larger porosity and smallest particle density that may be due to the presence of the oil in the spent zeolites.

Organic substances produce additional gases during the sintering process, which contribute to the formation of pore beads and the creation of more porous structure of the aggregate [56]. However, similarly low densities (0.7–0.9) were achieved by Volland and Brötz [3] for sand sludge LWAs admixed with 20–40% of heulanditic zeolite rock. Such low bulk densities (from 0.95 to 0.7) were also achieved by Mun [57] for LWAs admixed with different doses of a sewage sludge. Kourti and Cheeseman [26] found that sintering 60:40 lignite coal fly ash with waste glass mixes produced LWAs with a mean density of $1.35 \text{ g} \cdot \text{cm}^{-3}$, thereby suggesting that heat treatment of organic material containing substrates gives smaller bulk densities and higher porosities of the resulting LWAs than using organic-derived fly ashes coming from similar organic material. It is worth noting that practically the same bulk densities of the aggregates are measured by mercury intrusion and from the LWA volume and mass (Table 2), indicating that the amount of very fine pores being unavailable for mercury is very small in all LWAs studied. The solid phase (skeletal) density is the highest for CLAY and the smallest for CLIN aggregates. It could be possible that the presence of residual carbon formed from no oxidized oil additions diminished the solid phase density; however, no carbon could be detected in the LWAs studied. The finest close pores are possibly responsible for the above effect. The fractal dimensions calculated from microtomography data are rather high, indicating the complex pore buildup, of which the least diversified is found in the CLAY aggregate. All microtomography fractal dimensions fall within the range between 2 and 3, therefore it is likely that microtomography provides a more realistic picture of the LWAs' fractal pore structure than MIP. This may be due either to the application of the spherical pore model for microtomography data elaboration (instead of cylindrical pore spaces model in MIP) or more probably to a failure of the MIP application in describing LWAs' pore size distribution.

4. Conclusions

Although the addition of spent zeolite sorbents increased the amount of the amorphous glassy phase in the LWAs, their mineral composition stayed intact, as evidenced by the XRD results. The addition of spent zeolites has fostered a decrease in the particle density, which in turn has involved a decrease in the mechanical resistance. A decrease in water absorption also occurred. The pore structure of LWAs prepared from a clay deposit was strongly modified by the addition of spent zeolites, depending on the composition of the starting mixture. All the methods applied measured the same tendencies of changes in pore volumes and porosities of LWAs due to the addition of spent zeolites. The porosity of the LWAs prepared from a clay deposit was the lowest and the addition of spent NaP1 resulted in the highest porosity of the obtained LWAs. An increase in porosity may also be connected with the amount of the oil present within the added zeolites: with more oil addition the more porous structure is formed. Changes in the average pore radius measured by microtomography and MIP did not run parallel with the pore volume changes. Only the dominant pore radius measured by SEM increased to a similar degree as the porosity.

The reuse (addition) of the spent zeolitic sorbents containing petroleum waste to produce LWAs is a novel method dedicated to this kind of waste utilization. Furthermore, it leads to very advantageous properties of the resulting LWAs (high porosity, low water sorption, enough mechanical resistance, high freezing resistance), indicating their applicability in geotechnics, building construction and agriculture.

Acknowledgments: This research was financed within the statutory funds No. S12/II/B/2017. Funding Body: Ministry of Science and Higher Education (Republic of Poland).

Author Contributions: Wojciech Franus conceived and designed the experiment, performed and analysed the XRD and SEM data, and prepared the manuscript. Grzegorz Jozefaciuk performed and interpreted MIP and MT data, formulated main conclusions and translated the manuscript. Lidia Bandura prepared the ceramic materials, collected literature, took part in manuscript preparation, and provided manuscript formatting. Małgorzata Franus prepared the ceramic materials, performed physical and mechanical properties measurements, provided XRD and SEM interpretation, prepared the manuscript.

References

1. Sokolova, S.N.; Vereshagin, V.I. Lightweight granular material from zeolite rocks with different additives. *Constr. Build. Mater.* **2010**, *24*, 625–629. [CrossRef]
2. Riley, C.M. Relation of chemical properties to the bloating of clays. *J. Am. Ceram. Soc.* **1951**, *41*, 74–80. [CrossRef]
3. Volland, S.; Brötz, J. Lightweight aggregates produced from sand sludge and zeolitic rocks. *Constr. Build. Mater.* **2015**, *85*, 22–29. [CrossRef]
4. González-Corrochano, B.; Alonso-Azcárate, J.; Rodas, M. Characterization of lightweight aggregates manufactured from washing aggregate sludge and fly ash. *Resour. Conserv. Recycl.* **2009**, *53*, 571–581. [CrossRef]
5. González-Corrochano, B.; Alonso-Azcárate, J.; Rodas, M.; Luque, F.J.; Barrenechea, J.F. Microstructure and mineralogy of lightweight aggregates produced from washing aggregate sludge, fly ash and used motor oil. *Cem. Concr. Compos.* **2010**, *32*, 694–707. [CrossRef]
6. Fronczyk, J.; Radziemska, M.; Mazur, Z. Copper removal from contaminated groundwater using natural and engineered limestone sand in permeable reactive barriers. *Fresenius Environ. Bull.* **2015**, *24*, 228–234.
7. Topçu, I.B.; Işikdağ, B. Effect of expanded perlite aggregate on the properties of lightweight concrete. *J. Mater. Process. Technol.* **2008**, *204*, 34–38. [CrossRef]
8. Mouli, M.; Khelafi, H. Performance characteristics of lightweight aggregate concrete containing natural pozzolan. *Build. Environ.* **2008**, *43*, 31–36. [CrossRef]
9. Kockal, N.U.; Ozturan, T. Characteristics of lightweight fly ash aggregates produced with different binders and heat treatments. *Cem. Concr. Compos.* **2011**, *33*, 61–67. [CrossRef]
10. Korat, L.; Ducman, V.; Legat, A.; Mirtič, B. Characterisation of the pore-forming process in lightweight aggregate based on silica sludge by means of X-ray micro-tomography (micro-CT) and mercury intrusion porosimetry (MIP). *Ceram. Int.* **2013**, *39*, 6997–7005. [CrossRef]
11. Fragoulis, D.; Stamatakis, M.G.; Chaniotakis, E.; Columbus, G. Characterization of lightweight aggregates produced with clayey diatomite rocks originating from Greece. *Mater. Charact.* **2004**, *53*, 307–316. [CrossRef]
12. Sengul, O.; Azizi, S.; Karaosmanoglu, F.; Tasdemir, M.A. Effect of expanded perlite on the mechanical properties and thermal conductivity of lightweight concrete. *Energy Build.* **2011**, *43*, 671–676. [CrossRef]
13. González-Corrochano, B.; Alonso-Azcárate, J.; Rodas, M.; Barrenechea, J.F.; Luque, F.J. Microstructure and mineralogy of lightweight aggregates manufactured from mining and industrial wastes. *Constr. Build. Mater.* **2011**, *25*, 3591–3602. [CrossRef]
14. González-Corrochano, B.; Alonso-Azcárate, J.; Rodas, M. Effect of prefiring and firing dwell times on the properties of artificial lightweight aggregates. *Constr. Build. Mater.* **2014**, *53*, 91–101. [CrossRef]
15. Wei, Y.L.; Lin, C.Y.; Ko, K.W.; Wang, H.P. Preparation of low water-sorption lightweight aggregates from harbor sediment added with waste glass. *Mar. Pollut. Bull.* **2011**, *63*, 135–140. [CrossRef] [PubMed]
16. Go, C.G.; Tang, J.R.; Chi, J.H.; Chen, C.T.; Huang, Y.L. Fire-resistance property of reinforced lightweight aggregate concrete wall. *Constr. Build. Mater.* **2012**, *30*, 725–733. [CrossRef]
17. Ducman, V.; Mirtic, B. The applicability of different waste materials for the production of lightweight aggregates. *Waste Manag.* **2009**, *29*, 2361–2368. [CrossRef] [PubMed]
18. Moreira, A.; António, J.; Tadeu, A. Lightweight screed containing cork granules: Mechanical and hygrothermal characterization. *Cem. Concr. Compos.* **2014**, *49*, 1–8. [CrossRef]
19. Sarabèr, A.; Overhof, R.; Green, T.; Pels, J. Artificial lightweight aggregates as utilization for future ashes—A case study. *Waste Manag.* **2012**, *32*, 144–152. [CrossRef] [PubMed]
20. Quina, M.J.; Bordado, J.M.; Quinta-Ferreira, R.M. Recycling of air pollution control residues from municipal solid waste incineration into lightweight aggregates. *Waste Manag.* **2014**, *34*, 430–438. [CrossRef] [PubMed]
21. Franus, M.; Barnat-Hunek, D. Analysis of physical and mechanical properties of lightweight aggregate modified with sewage sludge. *Proc. ECOpole* **2015**, *9*, 33–39.
22. Franus, M.; Barnat-Hunek, D.; Wdowin, M. Utilization of sewage sludge in the manufacture of lightweight aggregate. *Environ. Monit. Assess.* **2016**, *188*, 10–23. [CrossRef] [PubMed]
23. Suchorab, Z.; Barnat-Hunek, D.; Franus, M.; Łagód, G. Mechanical and physical properties of hydrophobized lightweight aggregate concrete with sewage sludge. *Materials* **2016**, *9*, 317. [CrossRef]

24. Cheeseman, C.R.; Makinde, A.; Bethanis, S. Properties of lightweight aggregate produced by rapid sintering of incinerator bottom ash. *Resour. Conserv. Recycl.* **2005**, *43*, 147–162. [CrossRef]

25. Anagnostopoulos, I.M.; Stivanakis, V.E. Utilization of lignite power generation residues for the production of lightweight aggregates. *J. Hazard. Mater.* **2009**, *163*, 329–336. [CrossRef] [PubMed]

26. Kourti, I.; Cheeseman, C.R. Properties and microstructure of lightweight aggregate produced from lignite coal fly ash and recycled glass. *Resour. Conserv. Recycl.* **2010**, *54*, 769–775. [CrossRef]

27. Verbinnen, B.; Block, C.; van Caneghem, J.; Vandecasteele, C. Recycling of spent adsorbents for oxyanions and heavy metal ions in the production of ceramics. *Waste Manag.* **2015**, *45*, 407–411. [CrossRef] [PubMed]

28. Franus, W.; Franus, M.; Latosińska, J.; Wójcik, R. The use of spent glauconite in lightweight aggregate production. *Bol. Soc. Esp. Cerám. Vidr.* **2011**, *50*, 193–200. [CrossRef]

29. González-Corrochano, B.; Alonso-Azcárate, J.; Rodas, M. Effect of thermal treatment on the retention of chemical elements in the structure of lightweight aggregates manufactured from contaminated mine soil and fly ash. *Constr. Build. Mater.* **2012**, *35*, 497–507. [CrossRef]

30. Misaelides, P. Application of natural zeolites in environmental remediation: A short review. *Microporous Mesoporous Mater.* **2011**, *144*, 15–18. [CrossRef]

31. Woszuk, A.; Franus, W. Properties of the Warm Mix Asphalt involving clinoptilolite and Na-P1 zeolite additives. *Constr. Build. Mater.* **2016**, *114*, 556–563. [CrossRef]

32. Wdowin, M.; Franus, W.; Panek, R. Preliminary results of usage possibilities of carbonate and zeolitic sorbents in CO_2 capture. *Fresenius Environ. Bull.* **2012**, *21*, 3726–3734.

33. Radziemska, M.; Fronczyk, J. Level and contamination assessment of soil along an expressway in an ecologically valuable Area in central Poland. *Int. J. Environ. Res. Public Health* **2015**, *12*, 13372–13387. [CrossRef] [PubMed]

34. Bandura, L.; Franus, M.; Józefaciuk, G.; Franus, W. Synthetic zeolites from fly ash as effective mineral sorbents for land-based petroleum spills cleanup. *Fuel* **2015**, *147*, 100–107. [CrossRef]

35. Franus, M.; Wdowin, M.; Bandura, L.; Franus, W. Removal of environmental pollutions using zeolites from fly ash: A review. *Fresenius Environ. Bull.* **2015**, *24*, 854–866.

36. Bandura, L.; Franus, M.; Panek, R.; Woszuk, A.; Franus, W. Characterization of zeolites and their use as adsorbents of petroleum substances. *Przemysl Chemiczny* **2015**, *94*, 323–327.

37. Bandura, L.; Panek, R.; Rotko, M.; Franus, W. Synthetic zeolites from fly ash for an effective trapping of BTX in gas stream. *Microporous Mesoporous Mater.* **2016**, *223*, 1–9. [CrossRef]

38. Szala, B.; Bajda, T.; Matusik, J.; Zięba, K.; Kijak, B. BTX sorption on Na-P1 organo-zeolite as a process controlled by the amount of adsorbed HDTMA. *Microporous Mesoporous Mater.* **2015**, *202*, 115–123. [CrossRef]

39. Muir, B.; Bajda, T. Organically modified zeolites in petroleum compounds spill cleanup—Production, efficiency, utilization. *Fuel Process. Technol.* **2016**, *149*, 153–162. [CrossRef]

40. Franus, W.; Dudek, K. Clay minerals and clinoptilolite of Variegated Shales Formation of the Skole Unit, Polish Flysch Carpathians. *Geol. Carpathica* **1999**, *50*, 23–24.

41. Wdowin, M.; Franus, M.; Panek, R.; Badura, L.; Franus, W. The conversion technology of fly ash into zeolites. *Clean Technol. Environ. Policy* **2014**, *16*, 1217–1223. [CrossRef]

42. American Society for Testing and Materials. *Standard Test Method for Liquid Limit, Plastic Limit, and Plasticity Index of Soils*; ASTM D 4318; American Society for Testing and Materials: West Conshohocken, PH, USA, 2010.

43. BSI Standards Publication. *Test for Mechanical and Physical Properties of Aggregates. Part 6: Determination of Particle Density and Water Absorption*; BS EN 1097-6:2013; British Standards Institute: London, UK, 2013.

44. European Committee for Standardization. *Lightweight aggregates—Part 1: Lightweight Aggregates for Concrete, Mortar and Grout*; EN 13055-1/AC; European Committee for Standardization: Brussels, Belgium, 2004.

45. European Committee for Standardization. *Test for Thermal and Weathering Properties of Aggregates. Part 2: Magnesium Sulfate Test*; EN 12697-2:1999; European Committee for Standardization: Brussels, Belgium, 1999.

46. Ryżak, M.; Bieganowski, A. Determination of particle size distrubution of soil using laser diffraction-comparison with areometric method. *Int. Agrophys.* **2010**, *24*, 177–181.

47. Dong, H.; Blunt, M.J. Pore-network extraction from micro-computerized-tomography images. *Phys. Rev. E.* **2009**, *80*, 1–11. [CrossRef] [PubMed]

48. Sridharan, A.; Venkatappa Rao, G. Pore size distribution of soils from mercury intrusion porosimetry data. *Soil Sci. Soc. Am. Proc.* **1972**, *36*, 980–981. [CrossRef]

49. Pachepsky, Y.A.; Polubesova, T.A.; Hajnos, M.; Sokolowska, Z.; Jozefaciuk, G. Fractal parameters of pore surface area as influences by simulated soil degradation. *Soil Sci. Soc. Am. J.* **1995**, *59*, 68–75. [CrossRef]
50. Yokoya, N.; Yamamoto, K.; Funakubo, N. Fractal-based analysis and interpolation of 3D natural surface shapes and their application to terrain modeling. *Comput. Vis. Graph. Image Process.* **1989**, *46*, 284–302. [CrossRef]
51. Vogel, H.J.; Roth, K. Quantitative morphology and network representation of soil pore structure. *Adv. Water Resour.* **2001**, *24*, 233–242. [CrossRef]
52. Lee, W.E.; Souza, G.P.; McConville, C.J.; Tarvornpanich, T.; Iqbal, Y. Mullite formation in clays and clay-derived vitreous ceramics. *J. Eur. Ceram. Soc.* **2008**, *28*, 465–471. [CrossRef]
53. Hung, M.-F.; Hwang, C.-L. Study of fine sediments for making lightweight aggregate. *Waste Manag. Res.* **2007**, *25*, 449–456. [CrossRef] [PubMed]
54. Gunning, P.J.; Hills, C.D.; Carey, P.J. Production of lightweight aggregate from industrial waste and carbon dioxide. *Waste Manag.* **2009**, *29*, 2722–2728. [CrossRef] [PubMed]
55. Stegemann, J.A.; Côté, P.L. A proposed protocol for evaluation of solidified wastes. *Sci. Total Environ.* **1996**, *178*, 103–110. [CrossRef]
56. Chang, F.C.; Lo, S.L.; Lee, M.Y.; Ko, C.H.; Lin, J.D.; Huang, S.C.; Wang, C.F. Leachability of metals from sludge-based artificial lightweight aggregate. *J. Hazard. Mater.* **2007**, *146*, 98–105.
57. Mun, K.J. Development and tests of lightweight aggregate using sewage sludge for nonstructural concrete. *Constr. Build. Mater.* **2007**, *21*, 1583–1588. [CrossRef]

minerals

MDPI

Article

Strength Reduction of Coal Pillar after CO_2 Sequestration in Abandoned Coal Mines

Qiuhao Du [1,†], Xiaoli Liu [1,2,*,†], Enzhi Wang [1,2] and Sijing Wang [3]

1   State Key Laboratory of Hydro-Science and Engineering, Tsinghua University, Beijing 100084, China;
    duqiuhao1@163.com (Q.D.); nzwang@tsinghua.edu.cn (E.W.)
2   Sanjiangyuan Collaborative Innovation Center, Tsinghua University, Beijing 100084, China
3   Institute of Geology and Geophysics of the Chinese Academy of Sciences, Beijing 100029, China;
    wangsijing@126.com
*   Correspondence: xiaoli.liu@tsinghua.edu.cn; Tel.: +86-10-6279-4910; Fax: +86-10-6278-2159
†   These authors contributed equally to this work.

Academic Editors: Annalisa Martucci and Giuseppe Cruciani
Received: 3 November 2016; Accepted: 4 February 2017; Published: 17 February 2017

Abstract: CO_2 geosequestration is currently considered to be the most effective and economical method to dispose of artificial greenhouse gases. There are a large number of coal mines that will be scrapped, and some of them are located in deep formations in China. CO_2 storage in abandoned coal mines will be a potential option for greenhouse gas disposal. However, CO_2 trapping in deep coal pillars would induce swelling effects of coal matrix. Adsorption-induced swelling not only modifies the volume and permeability of coal mass, but also causes the basic physical and mechanical properties changing, such as elastic modulus and Poisson ratio. It eventually results in some reduction in pillar strength. Based on the fractional swelling as a function of time and different loading pressure steps, the relationship between volumetric stress and adsorption pressure increment is acquired. Eventually, this paper presents a theory model to analyze the pillar strength reduction after CO_2 adsorption. The model provides a method to quantitatively describe the interrelation of volumetric strain, swelling stress, and mechanical strength reduction after gas adsorption under the condition of step-by-step pressure loading and the non-Langmuir isothermal model. The model might have a significantly important implication for predicting the swelling stress and mechanical behaviors of coal pillars during CO_2 sequestration in abandoned coal mines.

Keywords: CO_2 sequestration; abandoned coal mine; adsorption; swelling effect; strength reduction

1. Introduction

Greenhouse gas emissions are the most important contributor to global climate change. Among all kinds of greenhouse gass, the contribution rate of CO_2 to greenhouse efficiency was 63% [1]. According to statistics, fossil fuel combustion and industrial emissions of CO_2 accounted for about 78% of total CO_2 emission in the ten years from 2000 to 2010 [2]. Currently, CO_2 storage in oil and gas fields, brine water, deep unmined coal seams and deep sea are considered to be effective and practical ways to reduce atmospheric CO_2 level, helping to slow global climate change and temperature rise trends, in the event that fossil fuels remain in use as a primary energy source. In the 2013 technology roadmap, the international energy agency (IEA) proposed an integrated approach to drop greenhouse gas emission by reducing the use of fossil fuels, improving energy efficiency, implementing new energy sources and carbon capture and sequestration (CCS) technology [3]. In addition, with the successful exploitation of coalbed methane and shale gas, its considerable economic benefit prompted many countries to begin in order to regard shale gas, coalbed methane, as alternative unconventional energy. Many scholars have proposed using CO_2 to enhanced coalbed methane (ECBM) production, and the presence of CO_2 will mechanically weaken the coal and thus create fractures, helping to increase the permeability, improve the coalbed methane production yield and simultaneously sequestrate CO_2 [4–8]. In addition, for the residual space volume constituted by goaf areas and principle infrastructures in abandoned coal mines, some researchers proposed CO_2 sequestration in abandoned coal mines following the example of natural gas storage in it [9–12], which will be a potential option for CO_2 disposal because China has a large number of scrapped coal mines, and 541 key coal mines will be gradually closed by 2020. CO_2 can be stored in abandoned coal mines in three states: adsorbed on the remaining coal, free in empty space or dissolved in mine water.

The adsorption of CO_2 in coal can result in coal matrix swelling due to the fact that it has the highest adsorption potential compared with other fluids such as CH_4 and N_2. Currently, two mechanisms are applied to explain the adsorption-induced swelling in coal. On the one hand, several authors have widely researched the polymer structure, degree of cross-linking, three-dimensional polymeric network structures, as well as flexibility characteristics of coal macromolecules from the perspective of chemistry and molecules, and consider that the lower molecular-weight solvent, such as CO_2 and CH_4, can enter the macromolecule cross-linked polymer mesh, causing the coal matrix macromolecular structure rearrangement, resulting in swelling [13–17]. One the other hand, some scholars attribute the swelling being due to the formation of microfractures as the result of different pore systems, maceral components and mineral stiffness [18–21]. Hol et al. [22] considered CO_2 induced both reversible (i.e., adsorption-induced swelling and elastic compression) and irreversible (i.e., adsorption-induced microfracturing) strains under unconfined conditions.

Both ECBM and CO_2 sequestration in coal seams are concerned with the coal reservoir permeability behavior, the adsorption-induced swelling of coal matrix can compress the pore space and cleat system to result in the distinct decrease of the permeability of coal mass, and several models have been proposed from the consideration of effective stress, cleat volume compressibility, gas sorption-induced strain effect as well as pressure pulse decay [23–29]. Ranjith et al. [30] developed a triaxial equipment to study the gas fluid flow and found that coal mass permeability for CO_2 decreased largely with the increase of effective stress than that of N_2, due to the matrix swelling by CO_2 adsorption in coal. In addition, Verma and Sirvaiya [31] utilized the artificial neural network (ANN) to predict the Langmuir volume and pressure constants during CO_2 adsorption in coal, and that the ANN method was more accurate than other models in their study.

The effects of CO_2 adsorption on the strength of coal have been studied widely [32–34]. According to Gibbs' theory, when a more reactive, higher-chemical-energy adsorbate is used to displace the original adsorbate in solid adorbent, the surface energy of rock mass would reduce, which can lead to some reduction in initiation tension stress for fracturing and eventually the coal becomes more prone to damage. In addition, considering the thermodynamics of adsorption of gases in porous solids, the changes in surface energy at the interface between the gas adsorbate and solid adsorbent result in swelling through the conversion between surface free energy and elastic strain energy [35,36]. Based on the theory and experiment proposed by Meyers, a theoretical model was derived by Pan and Connell [37] through the energy balance approach. Hol et al. [38] developed a thermodynamic model based on statistical mechanics, and the model combined adsorption in a stress-supporting solid with the poroelastic to derive the relationship of stress–strain-sorption of coal under unconfined swelling condition. Liu et al. [39] revised the model derived by Hol et al. [38], and a corrected expression was obtained based on both statistical mechanics and kinetic approaches. Furthermore, Hol et al. [40] found the apparent bulk modulus determined for CO_2-equilibrated state was approximately 25% lower compared to the evacuated state through experiment data analysis. Ranjith et al. [32] studied the crack closure, crack initiation and crack damage of coal subjected to saturation with CO_2. Ranjith and Perera [41] considered the effects of the cheat system on strength reduction of coal after CO_2 adsorption. Perera et al. [42] experimented with adsorption of gaseous and super-critical CO_2 on bituminous coal from the Southern Sydney Basin, Australia, and studied the mechanical properties of coal sample before and after adsorption. The results showed that, compared with the natural state uniaxial compressive strength (UCS), the gaseous CO_2 saturation reduced UCS by 53% and elastic modulus by 36% using gas saturation pressure of 6 MPa. However, the supercritical CO_2 saturation reduced UCS by 79% and elastic modulus by 74% using super-critical saturation pressure of 8 MPa. It is shown that the phase of CO_2 has a significant effect on physical properties of coal, the adsorption capacity, swelling effect and strength parameters.

Coal is a discontinuous structure comprised by many natural fractures and cleats (Figure 1), and coal seams are normally conceptualized by a matchstick model (Figure 2). After the colliery was scrapped, leaving pillars with a large number of cracks by mining action on both sides of the goaf. Meanwhile, CO_2 adsorption causes the coal mass to break down along the cleat system easily due to the fact that the adsorption mainly affects the cohesion of coal, and the reduction of cohesive force leads to the apparent plastic deformation areas. However, for the internal friction angle, it decreases to a certain extent and no longer keeps changes. Pillars with a large number of cleats and fractures act as sealing walls when CO_2 is stored in goafs and drifts. It means that the stability of the pillar decides the safety and sealing of CO_2 sequestration. Based on the above discussions and conclusions, it is significantly important to discuss the strength reduction of the coal pillar after CO_2 injection in the abandoned coal mines. In this paper, we focus our attention on strength reduction of the pillar when CO_2 sequestration in abandoned coal mines, through the relationship of sorption-strain based on unconfined conditions, and swelling-stress under uniaxial conditions. Finally, a strength reduction model is proposed to qualitatively understand the effects of CO_2 adsorption on the strength and failure mechanics of coal pillar in abandoned coal mines.

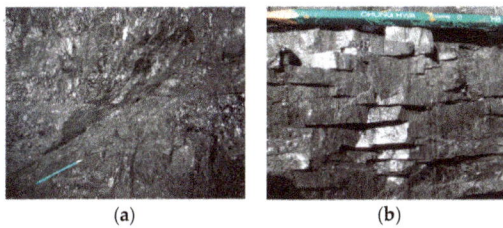

(a)   (b)

Figure 1. (a) Fractures, Chengzhuang Mine (Reproduced with permission from [43]); (b) cleats, Chengzhuang Mine (Reproduced with permission from [44]).

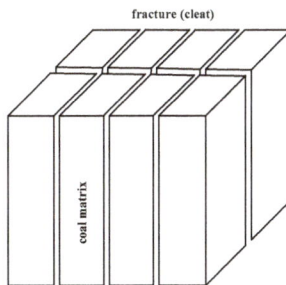

Figure 2. A matchstick model of a coal seam.

2. Theoretical Model

2.1. Adsorption-Induced Swelling Strain

CO_2 adsorption-induced volume strain has been studied by many researchers under unconfined conditions or uniaxial strain conditions [28,38,42]. In this section, we consider the adsorption-induced strain from the perspective of unconfined conditions. It is easy to measure the volumetric strain value of test coal samples when carrying out a gas adsorption experiment in the laboratory under unconfined conditions. As shown in Figure 3, the sample is a cube with the side length l. Two lengths are measured as the base values for volume calculation. One located at parallel (l_{pa}) to its bedding plane, and the other is perpendicular (l_{pe}). The parallel and perpendicular displacements were Δl_{pa} and Δl_{pe}, respectively.

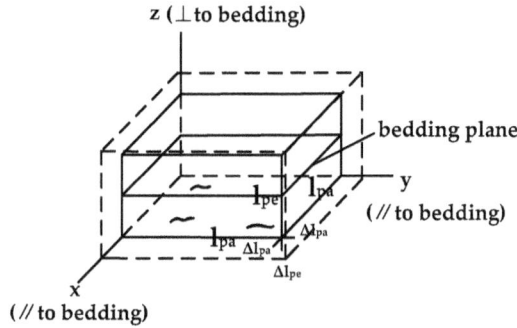

Figure 3. Sketch of original coal sample and swelling.

For the sake of convenience in the computation, it was assumed that parallel lengths and displacements of each block were equal and the perpendicular lengths and displacements were also equal. In this paper, the hypothesis that the coal mass satisfies the characteristics of isotropic and homogeneous is assumed. The reference value of volume was measured at the vacuum (initial volume V_0), and the volume increment was calculated based on the reference value with the pressure increasing step-by-step.

Here, according to the hypothesis and illustration above, the volume, as a function of time, can be written as:

$$V(t) = V_0 + \Delta V(t) = [l_{pa} + \Delta l_{pa}(t)] \times [l_{pa} + \Delta l_{pa}(t)] \times [l_{pe} + \Delta l_{pe}(t)] \tag{1}$$

$$l_{pa} = l_{pe}, \Delta l_{pa}(t) = \Delta l_{pe}(t) \tag{2}$$

At adsorption time t, the swelling of coal as a function of time is:

$$Q(t) = \frac{V(t) - V_0}{V_0} \tag{3}$$

The swelling before ith adsorption is $Q(t)_{i-1}$ (the swelling at the end of time exposure to P_{i-1}). At the ith adsorption pressure step, $Q(t)_i$ is the swelling at the end of time exposure to P_i. The fractional swelling increment is $q_i(t)$ during the coal sample is exposed to P_i [45]:

$$q_i(t) = \frac{\frac{V(t)-V_0}{V_0} - Q_{i-1}}{Q_i - Q_{i-1}} \tag{4}$$

where $q_i(t)$ is the volume strain change of the ith pressure step, which ranges from 0 to 1. $q_i(t) = 0$ means that adsorption just recently occurs at the P_i pressure step, and the swelling increment instantaneously changes with the adsorption amount. $q_i(t) = 1$ indicates that the swelling has reached equilibrium at the given gas pressure P_i. In order to facilitate the use of the elastic mechanic theory, Equation (4) is converted to the form as follows:

$$q_i(t) = \Delta \varepsilon_{vi}(t) \tag{5}$$

Here, the $\Delta \varepsilon_{vi}(t)$ means the fractional volumetric swelling increment from P_{i-1} to P_i.

Except for swelling, the given adsorption pressure generates volume compression due to closure of fractures and cleats, and coal matrix solid is also compressed during the loading process. In this process, the elastic modulus and Poisson's ratio are not fixed values. Goodman [46] suggested that the strain variation by pressure independent action was:

$$\varepsilon_P = -\frac{P}{E_s}(1 - 2\nu_s) \tag{6}$$

Exposure to the given P_i pressure step, the fractional strain variation of the sample is $\Delta\varepsilon_{Pi}$ from P_{i-1} to P_i

$$\Delta\varepsilon_{P_i} = -\frac{\Delta P_i}{E_s}(1 - 2\nu_s) \tag{7}$$

where ΔP_i is the pressure increment from P_{i-1} to P_i. E_s is the elastic modulus of coal matrix solid, which is not equivalent to the Young's modulus (E_P) that takes the elasticity of micropores into account. The relationship between the elastic modulus E_s and the Young's modulus E_P can be expressed as [47]:

$$E_s = \frac{E_P(3\rho_s - 2\rho)}{\rho} \tag{8}$$

ρ_s and ρ are the density of the solid phase (skeletal density) and apparent density, respectively.

According to Bentz et al. [48], Poisson' ratio change can be expressed in Equation (9) after sorption

$$\nu = \nu_s + \frac{3(1 - \nu_s^2)(1 - 5\nu_s)\phi}{2(7 - 5\nu_s)} \tag{9}$$

where ν_s is the Poisson ratio of the solid frame, ranging from -1 to 0.5, ν is the effective Poisson ratio, and ϕ is the porosity. In the process of deduction, it was assumed that the pore shape was cylindrical and the pores were randomly distributed. The relationships of $E_P/E_s - \rho/\rho_s$ as well as $\nu - \nu_s$ based on the research data of Bentz et al. [48] are shown in Figure 4.

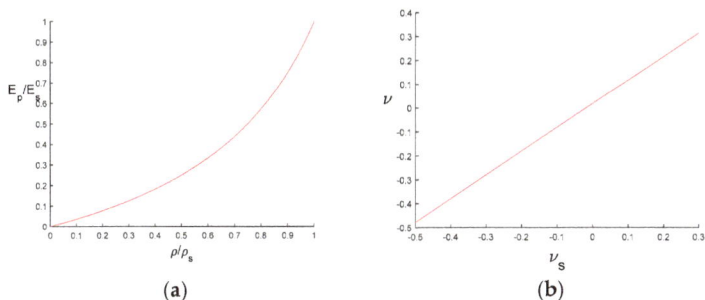

Figure 4. (a) The relationship of E_P/E_s and ρ/ρ_s; (b) the linear figure of ν against ν_s ($\phi = 0.07$) [48].

The fractional volumetric strain increment and the total volumetric strain is $\Delta\varepsilon_i$ and ε_i, respectively. Two parts are included, gas adsorption induced matrix swelling strain $\Delta\varepsilon_{vi}$, taken as positive, and adsorption pressure loading induced strain $\Delta\varepsilon_{Pi}$, taken as negative:

$$\Delta\varepsilon_i = \Delta\varepsilon_{vi} - \Delta\varepsilon_{Pi} \tag{10}$$

$$\varepsilon_i(t) = \sum_{i=1}^{n}(\Delta\varepsilon_{vi} - \Delta\varepsilon_{Pi}) = \sum_{i=1}^{n}\left[\frac{\frac{V(t)-V_0}{V_0} - Q_{i-1}}{Q_i - Q_{i-1}} + \frac{\Delta P_i}{E_s}(1 - 2\nu_s)\right] \tag{11}$$

Equation (11) is an expression of the total volumetric strain after equilibrium by pressure increasing step-by-step under unconfined conditions. This expression of adsorption-induced volumetric swelling strain considers the initial volume V_0 and the volume at the time of t. In addition, the changes of Young's modulus and Poisson ratio are also contained.

Under the hypothesis of isotropic and homogeneous, based on the relationship between linear strain and volumetric strain in elastic mechanics, and the linear swelling strain in the vertical direction can be expressed as follows:

$$\varepsilon_{iz}(t) = \frac{1}{3}\varepsilon_i(t) = \frac{1}{3}\sum_{i=1}^{n}[\frac{\frac{V(t)-V_0}{V_0} - Q_{i-1}}{Q_i - Q_{i-1}} + \frac{\Delta P_i}{E_s}(1 - 2v_s)] \tag{12}$$

2.2. Adsorption-Induced Stress under Uniaxial Conditions

After the mining work finished, only two parts of pillars and goafs are left in the original work face (Figure 5). Due to the overburden pressure and stress redistribution by mining activities, the pillar bears the pressure of overlying strata as well as mining stress and preforms uniaxial condition (Figure 6). If the CO_2 gas (fluid) is injected in goaf, this adsorption behavior will occur in the pillar. Under the uniaxial condition, adsorption-induced swelling strain occurs on the two sides adjacent to goafs and drifts, but the swelling strain of the direction perpendicular to bedding is inhibited, resulting in the swelling stress occurring in the vertical direction (z axial, Figure 3). The swelling stress can decrease crack initiation stress, resulting in the damage of the coal pillar. In this section, we consider the derivation of swelling stress in the direction perpendicular to bedding.

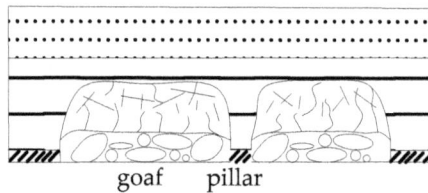

goaf   pillar

Figure 5. Schematic diagram of pillar and goaf in the abandoned coal mine.

Figure 6. Schematic diagram of the uniaxial condition of the pillar.

Based on the analysis of the section above, under uniaxial conditions, the constitutive relation of swelling stress and swelling strain in the vertical direction (z axial) follows Equation (13):

$$\sigma_{iz} = E_s \varepsilon_{iz} \tag{13}$$

Combining Equations (12)–(14), stress increment in the vertical direction in the process of exposure from P_{i-1} to P_i can be calculated by Formula (15). The swelling stress includes two parts. The first part is the adsorption-induced swelling stress, and the second is volume-compressed stress:

$$\Delta \varepsilon_{iz}(t) = \frac{1}{3}\Delta \varepsilon_i(t) \tag{14}$$

$$\Delta \sigma_{iz}(t) = E_s \Delta \varepsilon_{iz}(t) = \frac{E_s}{3}[\frac{\frac{V(t)-V_0}{V_0} - Q_{i-1}}{Q_i - Q_{i-1}} + \frac{\Delta P_i}{E_s}(1 - 2v_s)] \tag{15}$$

If, at the given nth pressure step, P_n, the pillar perfectly reaches adsorption equilibrium, and the fractional swelling variation does not change and the linear strain also remains constant. In this case, the swelling stress is the cumulative value of stress increments at overall adsorption pressure steps:

$$\sigma_{iz}(t) = \sum_{i=1}^{n}\Delta \sigma_{iz}(t) = \sum_{i=1}^{n}\frac{E_s}{3}[\frac{\frac{V(t)-V_0}{V_0} - Q_{i-1}}{Q_i - Q_{i-1}} + \frac{\Delta P_i}{E_s}(1 - 2v_s)] \tag{16}$$

where $\sigma_{iz}(t)$ is the total swelling stress considering the adsorption occurring in the pillar in the vertical direction where swelling strain is inhibited under the uniaxial condition. The swelling stress is a cumulative value by step-by-step pressures loading from 1st to nth steps.

2.3. Mechanical Strength Change after Gas Adsorption

The pillar coal matrix swells after CO_2 injection in abandoned coal mines, while it shrinks when gas is expelled from coal mass. Adsorption expansion will inevitably lead to some changes in the mechanical properties of the pillar, such as the above mentioned parameters, elastic modulus, Poisson's ratio, bulk modulus as well as shear modulus, due to these parameters being highly effected by adsorption-induced fracturing. All of these variations result in the decrease of pillar strength during CO_2 sequestration in abandoned coal mines. Hu et al. [49] demonstrated that the adsorption of gas exhibits dual effects on the physical properties of coal, mechanical (swelling stress) and non-mechanical (erosion effect) effects. The swelling effect leads to the decrease of the interaction between coal particles, while the erosion effect reduces the surface energy and lowers the surface tension of the coal mass. Gas adsorption mainly affects the cohesion, and the decrease of cohesion leads to obvious plastic deformation of coal, but, for the internal friction angle, it decreases to a certain extent and the hold over no longer changes. Hol et al. [40] verified that gas sorption can lead to a decrease in bulk modulus, while an increase in swelling caused the strain hysteresis to be oversized during the process of loading–unloading. Hagin and Zoback [50] compared the adsorption characteristics of CO_2 with that of helium, and found that the Young's modulus decreased after the CO_2 saturation adsorption. Simultaneously, the static bulk modulus reduced by an order of magnitude. Yang and Zoback [51] observed that CO_2 injection into coal samples resulted in the volume increase, and the coal sample became more viscous and less elastic. Ranjith and Perera [41] considered the effects of cleat density and direction on CO_2 adsorption-induced strength reduction.

On the basis of Griffith line elastic fracture theory, when cracks propagate, a portion of the elastic energy is converted into a solid surface energy. The fracture is confined in the extended critical state when the release rate of the elastic energy is equal to the increase rate of the surface energy. In the plane stress state, the critical stress is:

$$\sigma_c = \sqrt{\frac{2E\gamma}{\pi a}} \tag{17}$$

where γ is the surface energy per unit area, N/m; π is the surface tension change value from the vacuum to the given adsorption condition, N/m. In the case of the plane strain state, E is replaced by

$E - (1 - v^2)$. Furthermore, the relationship between the solid linear expansion strain and the change in surface tension is as follows:

$$\varepsilon = \lambda \pi = \lambda (\gamma_0 - \gamma_s) \tag{18}$$

where γ_0 and γ_s are the vacuum state and surface tension after gas adsorption, respectively, and λ can be described as Equation (19) [52]:

$$\lambda = \frac{2S\rho}{9K_s} \tag{19}$$

$$K_s = \frac{E_s}{3(1 - 2v_s)} \tag{20}$$

where S is the specific surface area of the coal sample, and K_s is the apparent modulus.

Equation (18) is rewritten as:

$$\gamma_s = \gamma_0 - \frac{\varepsilon_{iz}}{\lambda} \tag{21}$$

By Equation (17), the critical propagation stress at the end of the given pressure P_n for the nth adsorption step is expressed as function of time and swelling stress:

$$\sigma_{ci}(t) = \sqrt{\frac{2E_s(\gamma_0 - \frac{\varepsilon_{iz}(t)}{\lambda})}{\pi a}} = \sqrt{\frac{2E_s(\gamma_0 - \frac{\sum_{i=1}^{n} \Delta\sigma_{iz}(t)}{E_s\lambda})}{\pi a}} \tag{22}$$

It can be written as:

$$\sigma_{ci}^2(t) = \frac{2E_s\gamma_0}{\pi a} - \frac{2\sum_{i=1}^{n} \Delta\sigma_{iz}(t)}{\lambda \pi a} \tag{23}$$

The square of the critical strength in vacuum conditions is σ_{c0}^2, as the following

$$\sigma_{c0}^2 = \frac{2E_s\gamma_0}{\pi a} \tag{24}$$

At the end of exposure to the nth adsorption pressure step, the strength reduction rate of the pillar is the ratio of $\sigma_{ci}^2(t)$ to σ_{c0}^2

$$\left(\frac{\sigma_{ci}(t)}{\sigma_{c0}}\right)^2 = 1 - \frac{\sum_{i=1}^{n} \Delta\sigma_{iz}(t)}{E_s\lambda\gamma_0} \tag{25}$$

That is,

$$\left(\frac{\sigma_{ci}(t)}{\sigma_{c0}}\right)^2 = 1 - \frac{E_s \sum_{i=1}^{n} [\frac{\frac{V(t)-V_0}{V_0} - Q_{i-1}}{Q_i - Q_{i-1}} + \frac{\Delta P_i}{E_s}(1 - 2v_s)]}{2S\rho\gamma_0(1 - 2v_s)} \tag{26}$$

Equation (26) is the calculation formula of the strength reduction rate considering the swelling stress increment under the condition of gas adsorption at partial pressure loading step-by-step. The formula can be used to calculate the strength reduction value of gas adsorption conveniently. The curve of ratio $\sigma_{ci}^2(t)$ to σ_{c0}^2 is showed in Figure 7, and the strength reduction is nearly 17% after adsorption equilibrated based on the swelling strain data at pressure 2 MPa from [45]. Compared with other formulas of linear pressure adsorption models, Equation (14) is a progressive accumulation form that could be applied to calculate the swelling stress of nonlinear adsorption pressure loading–unloading. Moreover, in the case of re-adsorption after the coal sample having already swelled to a certain degree, Equations (16) and (26) are also helpful for determining the swelling stress fractional increment and strength reduction value.

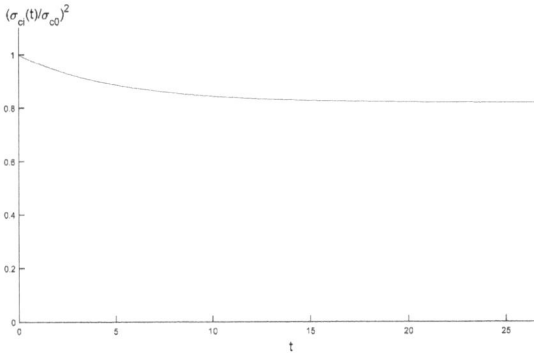

Figure 7. The reduction ratio curve of adsorption strength ($\sigma_{ci}^2(t)$) to original strength (σ_{c0}^2) over time. The swelling strain data are collected from [45].

3. Discussion

3.1. Swelling Strain and Swelling Stress

In this paper, we consider the volumetric swelling strain $\varepsilon_i(t)$ and isotropic linear strain ($\varepsilon_{iz}(t)$) in the z axial direction (Figure 3) as function of time after equilibrium by adsorption pressure increasing step-by-step under unconfined conditions. In connection with swelling strain, Hol et al. [38] and Liu et al. [39] considered thermodynamic models of gas adsorption and studied the effect of stress on the adsorption concentration of gas as well as sorption behavior. Their starting point is different from this paper based on directly volumetric changes. In the paper of Liu et al. [39], they established the relationship of internal energy, chemical potential, entropy change as well as stress–strain work on a single molecule of gas absorbed by the coal matrix cube. The strain was divided into mean extensional strain and deviatoric strain. In the deviation process of volumetric strain in this paper, it is easy to measure the strain increment without considering the thermodynamic process of adsorption. However, the coupling based on thermodynamic between stress–strain-sorption is significant important to understand the effect of pressure and temperature on adsorption in coal matrix. The swelling strain was divided into a reversible part and irreversible part under unconfined conditions [53]. Similarly, Wang et al. [54] divided strain into two parts at an isothermal condition. One is the mechanical deformation meeting the Hooke law stress–strain relationship and is calculated by the effective stress. The other is the deformation induced by gas adsorption or desorption. In Section 2, we do not mention it because not only an elastic strain but also an irreversible strain are converted to swelling stress considered from a macro perspective in the vertical direction under uniaxial loading.

In addition, based on the energy conservation law, and from the viewpoint of surface tension change, Wu et al. [55] and Bai et al. [56] utilized the principle that the expansion of the strain energy equals the surface tension work in order to establish the constitutive equation of swelling stress and swelling strain. Some researchers held the view that gas adsorption satisfies the Langmuir isothermal adsorption model (Figure 8), and considers the linear relationship between adsorption capacity and swelling strain. Although a large number of studies have shown that gas adsorption presents a single molecule arrangement structure that meets the Langmuir model, Lin [57] and Yu et al. [58] suggested that CO_2 adsorption on the coal mass can be described in a multi adsorbed layer model, i.e., the Brunauer-Emmett-Teller (BET) adsorption type model.

Langmuir isotherm adsorption model equation is given by:

$$q = \frac{abp_p}{1 + bp_p} \tag{27}$$

where q is the adsorbed amount of gas during adsorption reaching equilibrium, a is the gas limit adsorption capacity at the reference pressure, b is the adsorption equilibrium constant, and p_p is the pore pressure.

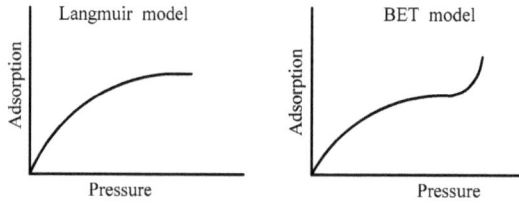

Figure 8. Two types of adsorption isotherms.

In the study of the acid gas CO_2 and H_2S storage in coal seam, Chikatamarla et al. [59] also calculated the frictional expansion strain by the Langmuir model:

$$V = \frac{V_L P}{P_L + P} \tag{28}$$

$$\varepsilon_V = \frac{\varepsilon_L P}{p + p_e} \tag{29}$$

where V, P, V_L, P_L are gas adsorption volume, adsorption pressure, Langmuir volume and Langmuir pressure, respectively. In addition, ε_V and ε_L are, respectively, the strain at given pressure and strain at infinite pressure, and p_e is the Langmuir pressure constant, which is equal to the pressure value when the strain at the pressure is half of the Langmuir maximum strain.

For the cylindrical specimen, according to the isotropic assumption of swelling/collapse in the process of adsorption/desorption, the volume strain is expressed as:

$$\varepsilon = \frac{\Delta V}{V} = \varepsilon_r^2 + 2\varepsilon_r + \varepsilon_a + \varepsilon_r^2 \varepsilon_a + 2\varepsilon_a \varepsilon_r \tag{30}$$

where V and ΔV are the initial volume and the volume change, respectively. ε_a and ε_r are, respectively, the axial strain and the radial strain.

Robertson and Christiansen [60] introduced the strain factor function to modify the Langmuir strain constant. The strain factor function is given by

$$s_f = \frac{p_{ob}}{p} [a + b(\frac{P_L}{\varepsilon_L V_r^2 \sqrt{\gamma}})] \tag{31}$$

where p_{ob} denotes the overburden pressure, a and b are empirical constants, V_r denotes the vitrinite reflectance of coal, and γ denotes the bulk density of the gas. p, ε_L and P_L are consistent with Equations (28) and (29).

The BET adsorption model is a generalization of the Langmuir model considering multi adsorbed molecule layers, which is expressed by [57,61]

$$q = \frac{V_m C P}{(P_0 - P)[1 + (C - 1)P/P_0]} \tag{32}$$

where P_0 is the saturation pressure of gas, P is the adsorption pressure, V_m is the maximum adsorption volume at the time the entire adsorbent surface is covered with a complete single molecular layer, and C is a constant related to the net heat of adsorption.

On the basis of the isotropic assumption and the Langmuir adsorption model, the above mentioned literature obtained the stress–strain relationship of swelling. All of those models follow a certain degree of representation in accordance with adsorption induced-swelling. However, they are not perfect because the Langmuir model has a very high accuracy and precision in the case of describing low-pressure gas adsorption. However, it is not suitable for high temperature gas adsorption, particularly, when CO_2 is sequestrated in deep unmined coal seams, where it is confined in states of high temperature and high pressure. Under the situation that CO_2 may exhibit a supercritical state that includes dual characteristics of gas and liquid, and the density of supercritical CO_2 fluid is close to that of the liquid, but the viscosity of which is similar to that of the gas. The diffusion coefficient of supercritical CO_2 is nearly one hundred times of that of the liquid.

Although this paper is also based on the isotropic hypothesis to establish the stress–strain relationship as a function of time and each adsorption pressure step, we do not consider the Langmuir isothermal adsorption model as the reference model but directly analyze it based on volume increment. Furthermore, other models only consider the initial state and the final equilibrium state and have exclusively been expressed in single integral steps, whilst the swelling increment of the prior swelled sample under re-loading is not taken into account.

3.2. Stress and Strength Reduction

Coupling relationship of swelling strains, swelling stressed and total gas uptake are affected by coal properties. Fractured coal possesses dual porosity system: (a) the cleat macroporosity system, and (b) the microporosity of coal matrix. Espinoza et al. [62,63] proposed a double porosity poromechanical model and considered the strains as a function of stresses, fracture pore pressure, and the pressure-dependent adsorption stress developed by the coal matrix.

Liu et al. [64] considered the seepage model of fracture–matrix interaction during coal deformation, and put forward the concept of internal swelling stress, σ_I (Figure 9). The strain is considered including two parts: the first part is the coal matrix strain induced by the internal swelling stress, and the second part is the volume strain of the fracture. Their work modified the form of effective stress, so that the corrected expression of effective stress can directly reflect its impact on the permeability. Nevertheless, the elastic modulus, Poisson's ratio as well as other parameters are variable rather than fixed values during the occurrence of coal matrix adsorption swelling. These variates are not taken into account in the derivation of theory and formula, which leads to some errors in the accuracy of the model.

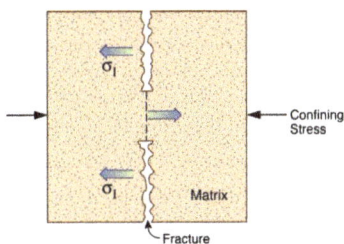

Figure 9. Graphical description of the internal swelling stress ([64]).

However, desorption differs from adsorption. Gas desorption can induce coal matrix shrinkage and stress relaxation. Espinoza et al. [65] studied the desorption-induced shear failure under zero-lateral strain condition and simulated the stress path far from the wellbore. Pore pressure in the coal cleats and desorption-induced variations of stresses jointly affect the stress path, i.e., poroelastic coefficient and variation changes of fluid pressure, as well as lateral stress decrease by desorption. The two part change effective stress of coal results in the stress path reaching the failure envelope. Wang et al. [66] considered gas desorption weakens coal by reducing effective stress. Desorption-induced shrinkages can relax lateral stress and aggravate shear failure of coal.

Ranjith et al. [32] considered CO_2 adsorption and mechanical behavior of coal, and held that peak strength could decrease when coal is saturated with CO_2. However, they did not comprehensively interpret the strength reduction mechanism. In this research, pillars are assumed to reach equilibrium on both sides of goaf after the mining working finished, and pillars are under the uniaxial condition. The coal pillar strength reduction is theoretically analyzed, and adsorption can increase the swelling stress on the faces of microfractures and cleats, resulting in effective stress decreasing. Under the condition of uniaxial conditions, if the loading pressure remains constant, and no swelling occurs, cracks could maintain the equilibrium state. However, when adsorption behaviors take place, the crack surface tension changes, and swelling stress emerges by swelling strain being inhibited. The stress needed to cause crack initiation is decreased, leading to coal pillars potentially being more prone to failure. The model proposed in this paper can describe the effect of swelling stress on strength clearly.

Coal is a solid that contains a large number of slit-like pores and fractures inter-connected by narrow capillary constriction and connected to the surface [67], which presents absolutely anisotropy and heterogeneity. Simultaneously, the adsorption induced-swelling also has anisotropic characteristics for the reason that most microfractures developed parallel to the bedding plate, approximately following maceral–maceral and bedding interfaces and swelling anisotropic [53]. In this paper, it is assumed that both coal and swelling have isotropic elastic properties for conveniently facilitating the application of existing theories and simplifying the process of calculation. From the perspective of CO_2 sequestration in abandoned coal mines, the strength change of coal pillars must be emphasized to prevent CO_2 leakage from goafs. To accurately predict the swelling stress in the vertical direction and strength reduction of pillars, much more research and many more approaches are still required.

4. Conclusions

This study has focused on volumetric swelling strain and strength reduction of pillars when CO_2 is stored in abandoned coal mines. The volumetric swelling strain is theoretically derived as a function of time by adsorption pressure increasing step-by-step under unconfined conditions. In connection with the conditions of coal pillars in abandoned coal mines, and a uniaxial loading model is proposed by simplifying the actual condition. Swelling strain in a direction perpendicular to bedding is inhibited when CO_2 adsorption is in the pillar. The effect of adsorption on pillar strength reduction is theoretically analyzed and deduced. Our findings can be summarized as follows:

1. There are a large number of coal mines that will be closed and some of them are located in deep formations in China. CO_2 storage in abandoned coal mines could be a potential option for greenhouse gas disposal.
2. The volume strain and swelling stress, as a function of time, and different loading pressure steps are deduced. Equation (15) is used to describe swelling stress considering coal has already had prior swelling deformation under the condition of step-by-step non-linear loading and a non-Langmuir isothermal model. The model presented in this paper is different from other models, in which only the initial state and the final equilibrium state are considered, and the incremental swelling process is neglected.
3. A theoretical model based on linear swelling stress–strain work is proposed to calculate the reduction ratio of coal pillar strength under uniaxial conditions. This theoretical model can be used to describe strength reduction during adsorption under adsorption pressure loading step-by-step.

Acknowledgments: The National Basic Research Program of China (Grant No. 2011CB013500), the National Key Research and Development Plan (Grant No. 2016YFC0501104), the National Natural Science Foundation of China (Grant No. U1361103, 51479094, 51379104), the National Natural Science Foundation Outstanding Youth Foundation (Grant No. 51522903), and the Open Research Fund Program of the State Key Laboratory of Hydroscience and Engineering (Grant 2013-KY-06, 2015-KY-04, 2016-KY-02) are gratefully acknowledged.

Author Contributions: Xiaoli Liu designed the research; Qiuhao Du performed the research; Enzhi Wang and Sijing Wang contributed reagents/materials/analysis tools; and Qiuhao Du wrote the paper. All authors read and approved the manuscript.

Conflicts of Interest: The authors declare no conflict of interest.

References

1. Zhang, F.; Zhou, H.; Lu, T.; Hu, D.W.; Sheng, Q.; Hu, Q.Z. Analysis of reservoir deformation and fluid transportation induced by injection of CO_2 into saline aquifer: (I) Two-phase flow-reservoir coupling model. *Rock Soil Mech.* **2014**, *35*, 2549–2554. (In Chinese)

2. ·Edenhofer, O. Mitigation of climate change IA models and WGIII: Lessons from IPCC AR5. In Proceedings of the 7th IAMC Meeting, Valencia, Spain, 23–25 October 2014.

3. International Energy Agency (IEA). *Four Energy Policies Can Keep the 2 °C Climate Goal Alive*; International Energy Agency (IEA): Paris, France, 2013.

4. Seomoon, H.; Lee, M.; Sung, W. Analysis of methane recovery through CO_2-N_2 mixed gas injection considering gas diffusion phenomenon in coal seam. *Energy Explor. Exploit.* **2016**, *34*, 661–675. [CrossRef]

5. He, L.; Shen, P.; Liao, X.; Li, F.; Gao, Q.; Wang, Z. Potential evaluation of CO_2 EOR and sequestration in Yanchang oilfield. *J. Energy Inst.* **2016**, *89*, 215–221. [CrossRef]

6. Chareonsuppanimit, P.; Mohammad, S.A.; Robinson, R.L.; Gasem, K.A.M. High-pressure adsorption of gases on shales: Measurements and modeling. *Int. J. Coal Geol.* **2012**, *95*, 34–46. [CrossRef]

7. Lu, Y.; Ao, X.; Tang, J.; Jia, Y.; Zhang, X.; Chen, Y. Swelling of shale in supercritical carbon dioxide. *J. Nat. Gas Sci. Eng.* **2016**, *30*, 268–275. [CrossRef]

8. Day, S.; Fry, R.; Sakurovs, R.; Weir, S. Swelling of coals by supercritical gases and its relationship to sorption. *Energy Fuels* **2010**, *24*, 2777–2783. [CrossRef]

9. Piessens, K.; Dusar, M. Feasibility of CO_2 sequestration in abandoned coal mines in Belgium. *Geol. Belg.* **2004**, *7*, 168–180.

10. Piessens, K.; Dusar, M. Integration of CO_2 sequestration and CO_2 geothermics in energy systems for abandoned coal mines. *Geol. Belg.* **2004**, *7*, 181–189.

11. Van Tongeren, P.; Dreesen, R. Residual space volumes in abandoned coal mines of the Belgian Campine basin and possibilities for use. *Geol. Belg.* **2004**, *7*, 157–164.

12. Jalili, P.; Saydam, S.; Cinar, Y. CO_2 storage in abandoned coal mines. In Proceedings of the 2011 Underground Coal Operators' Conference, Beijing, China, 8–9 January 2011.

13. Van Krevelen, D.W. *Coal-Typology, Chemistry, Physics, Constitution*; Elsevier: Amsterdam, The Netherlands, 1961.

14. Sanada, Y.; Honda, H. Swelling equilibrium of coal by pyridine at 25 degrees C. *Fuel* **1966**, *45*, 295.

15. Solomon, P.R.; Fletcher, T.H. Impact of coal pyrolysis on combustion. *Symp. (Int.) Combust.* **1994**, *25*, 463–474.

16. Walker, P.L.; Verma, S.K.; Rivera-Utrilla, J.; Khan, M.R. A direct measurement of expansion in coals and macerais induced by carbon dioxide and methanol. *Fuel* **1988**, *67*, 719–726. [CrossRef]

17. Karacan, C.Ö. Swelling-induced volumetric strains internal to a stressed coal associated with CO_2 sorption. *Int. J. Coal Geol.* **2007**, *72*, 209–220. [CrossRef]

18. Pan, Z.; Connell, L.D. Modelling of anisotropic coal swelling and its impact on permeability behaviour for primary and enhanced coalbed methane recovery. *Int. J. Coal Geol.* **2011**, *85*, 257–267. [CrossRef]

19. Feng, Z.; Zhou, D.; Zhao, Y.; Cai, T. Study on microstructural changes of coal after methane adsorption. *J. Nat. Gas Sci. Eng.* **2016**, *30*, 28–37. [CrossRef]

20. Shovkun, I.; Espinoza, D.N.; Ramos, M.J. Coupled reservoir simulation of geomechanics and fluid flow in organic-rich rocks: Impact of gas desorption and stress changes on permeability during depletion. In Proceedings of the 50th US Rock Mechanics/Geomechanics Symposium, Houston, TE, USA, 26–29 June 2016.

21. Heller, R.; Zoback, M. Adsorption of methane and carbon dioxide on gas shale and pure mineral samples. *J. Unconv. Oil Gas Resour.* **2014**, *8*, 14–24. [CrossRef]

22. Hol, S.; Spiers, C.J. Competition between adsorption-induced swelling and elastic compression of coal at CO_2 pressures up to 100 MPa. *J. Mech. Phys. Solids* **2012**, *60*, 1862–1882. [CrossRef]

23. Zang, J.; Wang, K. Gas sorption-induced coal swelling kinetics and its effects on coal permeability evolution: Model development and analysis. *Fuel* **2017**, *189*, 164–177. [CrossRef]

24. Feng, R.; Harpalani, S.; Pandey, R. Laboratory measurement of stress-dependent coal permeability using pulse-decay technique and flow modeling with gas depletion. *Fuel* **2016**, *177*, 76–86. [CrossRef]
25. Connell, L.D. A new interpretation of the response of coal permeability to changes in pore pressure, stress and matrix shrinkage. *Int. J. Coal Geol.* **2016**, *162*, 169–182. [CrossRef]
26. Zhang, L.; Zhang, C.; Tu, S.; Tu, H.; Wang, C. A study of directional permeability and gas injection to flush coal seam gas testing apparatus and method. *Transp. Porous Media* **2016**, *111*, 573–589. [CrossRef]
27. Connell, L.D.; Mazumder, S.; Sander, R.; Camilleri, M.; Pan, Z.; Heryanto, D. Laboratory characterisation of coal matrix shrinkage, cleat compressibility and the geomechanical properties determining reservoir permeability. *Fuel* **2016**, *165*, 499–512. [CrossRef]
28. Peng, Y.; Liu, J.; Pan, Z.; Connell, L.D.; Chen, Z.; Qu, H. Impact of coal matrix strains on the evolution of permeability. *Fuel* **2017**, *189*, 270–283. [CrossRef]
29. Jasinge, D.; Ranjith, P.G.; Choi, X.; Fernando, J. Investigation of the influence of coal swelling on permeability characteristics using natural brown coal and reconstituted brown coal specimens. *Energy* **2012**, *39*, 303–309. [CrossRef]
30. Ranjith, P.G.; Perera, M.S.A. A new triaxial apparatus to study the mechanical and fluid flow aspects of carbon dioxide sequestration in geological formations. *Fuel* **2011**, *90*, 2751–2759. [CrossRef]
31. Verma, A.K.; Sirvaiya, A. Comparative analysis of intelligent models for prediction of Langmuir constants for CO_2 adsorption of Gondwana coals in India. *Geomech. Geophys. Geo Energy Geo Resour.* **2016**, *2*, 97–109. [CrossRef]
32. Ranjith, P.G.; Jasinge, D.; Choi, S.K.; Mehic, M.; Shannon, B. The effect of CO_2 saturation on mechanical properties of Australian black coal using acoustic emission. *Fuel* **2010**, *89*, 2110–2117. [CrossRef]
33. Viete, D.R.; Ranjith, P.G. The mechanical behaviour of coal with respect to CO_2 sequestration in deep coal seams. *Fuel* **2007**, *86*, 2667–2671. [CrossRef]
34. Vishal, V.; Ranjith, P.G.; Singh, T.N. An experimental investigation on behaviour of coal under fluid saturation, using acoustic emission. *J. Nat. Gas Sci. Eng.* **2015**, *22*, 428–436. [CrossRef]
35. Myers, A.L. Thermodynamics of adsorption in porous materials. *AIChE J.* **2002**, *48*, 145–160. [CrossRef]
36. Myers, A.L.; Monson, P.A. Adsorption in porous materials at high pressure: Theory and experiment. *Langmuir* **2002**, *18*, 10261–10273. [CrossRef]
37. Pan, Z.; Connell, L.D. A theoretical model for gas adsorption-induced coal swelling. *Int. J. Coal Geol.* **2007**, *69*, 243–252. [CrossRef]
38. Hol, S.; Peach, C.J.; Spiers, C.J. Effect of 3-D stress state on adsorption of CO_2 by coal. *Int. J. Coal Geol.* **2012**, *93*, 1–15. [CrossRef]
39. Liu, J.; Spiers, C.J.; Peach, C.J.; Vidal-Gilbert, S. Effect of lithostatic stress on methane sorption by coal: Theory vs. experiment and implications for predicting in-situ coalbed methane content. *Int. J. Coal Geol.* **2016**, *167*, 48–64. [CrossRef]
40. Hol, S.; Gensterblum, Y.; Massarotto, P. Sorption and changes in bulk modulus of coal—Experimental evidence and governing mechanisms for CBM and ECBM applications. *Int. J. Coal Geol.* **2014**, *128*, 119–133. [CrossRef]
41. Ranjith, P.G.; Perera, M.S.A. Effects of cleat performance on strength reduction of coal in CO_2 sequestration. *Energy* **2012**, *45*, 1069–1075. [CrossRef]
42. Perera, M.S.A.; Ranjith, P.G.; Viete, D.R. Effects of gaseous and super-critical carbon dioxide saturation on the mechanical properties of bituminous coal from the Southern Sydney Basin. *Appl. Energy* **2013**, *110*, 73–81. [CrossRef]
43. Wang, S.; Hou, G.; Zhang, M.; Sun, Q. Analysis of the visible fracture system of coalseam in Chengzhuang Coalmine of Jincheng City, Shanxi Province. *Chin. Sci. Bull.* **2005**, *50*, 45–51. [CrossRef]
44. Liu, S.; Sang, S.; Liu, H.; Zhu, Q. Growth characteristics and genetic types of pores and fractures in a high-rank coal reservoir of the southern Qinshui basin. *Ore Geol. Rev.* **2015**, *64*, 140–151. [CrossRef]
45. Staib, G.; Sakurovs, R.; Gray, E.M.A. Kinetics of coal swelling in gases: Influence of gas pressure, gas type and coal type. *Int. J. Coal Geol.* **2014**, *132*, 117–122. [CrossRef]
46. Goodman, R.E. *Introduction to Rock Mechanics*; John Wiley & Sons: Hoboken, NY, USA, 1980.
47. Scherer, G.W. Dilatation of porous glass. *J. Am. Ceram. Soc.* **1986**, *69*, 473–480. [CrossRef]
48. Bentz, D.P.; Garboczi, E.J.; Quenard, D.A. Modelling drying shrinkage in reconstructed porous materials: Application to porous Vycor glass. *Model. Simul. Mater. Sci. Eng.* **1998**, *6*, 211. [CrossRef]

49. Hu, S.; Wang, E.; Li, X.; Bai, B. Effects of gas adsorption on mechanical properties and erosion mechanism of coal. *J. Nat. Gas Sci. Eng.* **2016**, *30*, 531–538. [CrossRef]

50. Hagin, P.N.; Zoback, M.D. Laboratory studies of the compressibility and permeability of low-rank coal samples from the Powder River Basin, Wyoming, USA. In Proceedings of the 44th US Rock Mechanics Symposium and 5th US-Canada Rock Mechanics Symposium, Salt Lake City, UT, USA, 27–30 June 2010.

51. Yang, Y.; Zoback, M.D. The effects of gas adsorption on swelling, visco-plastic creep and permeability of sub-bituminous coal. In Proceedings of the 45th U.S. Rock Mechanics/Geomechanics Symposium, San Francisco, CA, USA, 26–29 June 2011.

52. Adamson, A.W.; Gast, A.P. *Physical Chemistry of Surfaces*; Science Press: Beijing, China, 1984. (In Chinese)

53. Hol, S.; Spiers, C.J.; Peach, C.J. Microfracturing of coal due to interaction with CO_2 under unconfined conditions. *Fuel* **2012**, *97*, 569–584. [CrossRef]

54. Wang, G.X.; Massarotto, P.; Rudolph, V. An improved permeability model of coal for coalbed methane recovery and CO_2 geosequestration. *Int. J. Coal Geol.* **2009**, *77*, 127–136. [CrossRef]

55. Wu, S.Y.; Zhao, W. Analysis of effective stress in adsorbed methane-coal system. *Chin. J. Rock Mech. Eng.* **2005**, *24*, 1674–1678. (In Chinese)

56. Bai, B.; Li, X.C.; Liu, Y.F.; Fang, Z.M.; Wang, W. Preliminary theoretical study on impact on coal caused by interactions between CO_2 and coal. *Rock Soil Mech.* **2007**, *28*, 823–826. (In Chinese)

57. Lin, W. Gas Sorption and the Consequent Volumetric and Permeability Change of Coal. Ph.D. Thesis, Stanford University, Stanford, CA, USA, March 2010.

58. Yu, W.; Al-Shalabi, E.W.; Sepehrnoori, K. A sensitivity study of potential CO_2 injection for enhanced gas recovery in Barnett shale reservoirs. In Proceedings of the SPE Unconventional Resources Conference, The Woodlands, TX, USA, 1–3 April 2014.

59. Chikatamarla, L.; Cui, X.; Bustin, R.M. Implications of volumetric swelling/shrinkage of coal in sequestration of acid gases. In Proceedings of the International Coalbed Methane Symposium, Tuscaloosa, AL, USA, 3–7 May 2004.

60. Robertson, E.P.; Christiansen, R.L. Modeling laboratory permeability in coal using sorption-induced strain data. *SPE Reserv. Eval. Eng.* **2007**, *10*, 260–269. [CrossRef]

61. Sing, K.S.W. Reporting physisorption data for gas/solid systems with special reference to the determination of surface area and porosity (Recommendations 1984). *Pure Appl. Chem.* **1985**, *57*, 603–619. [CrossRef]

62. Espinoza, D.N.; Vandamme, M.; Pereira, J.M.; Dangla, P.; Vidal-Gilbert, S. Measurement and modeling of adsorptive-poromechanical properties of bituminous coal cores exposed to CO_2: Adsorption, swelling strains, swelling stresses and impact on fracture permeability. *Int. J. Coal Geol.* **2014**, *134*, 80–95. [CrossRef]

63. Espinoza, D.N.; Vandamme, M.; Dangla, P.; Pereira, J.M.; Vidal-Gilbert, S. Adsorptive-mechanical properties of reconstituted granular coal: Experimental characterization and poromechanical modeling. *Int. J. Coal Geol.* **2016**, *162*, 158–168. [CrossRef]

64. Liu, H.H.; Rutqvist, J. A new coal-permeability model: Internal swelling stress and fracture-matrix interaction. *Transp. Porous Media* **2010**, *82*, 157–171. [CrossRef]

65. Espinoza, D.N.; Pereira, J.M.; Vandamme, M.; Dangla, P.; Vidal-Gilbert, S. Desorption-induced shear failure of coal bed seams during gas depletion. *Int. J. Coal Geol.* **2015**, *137*, 142–151. [CrossRef]

66. Wang, S.; Elsworth, D.; Liu, J. Permeability evolution during progressive deformation of intact coal and implications for instability in underground coal seams. *Int. J. Rock Mech. Min. Sci.* **2013**, *58*, 34–45. [CrossRef]

67. Melnichenko, Y.B.; He, L.; Sakurovs, R.; Kholodenko, A.L.; Blach, T.; Mastalerz, M.; Andrzej, P.; Radliński, A.; Cheng, G.; Mildner, D.F.R. Accessibility of pores in coal to methane and carbon dioxide. *Fuel* **2012**, *91*, 200–208. [CrossRef]

minerals

MDPI

Article

Temperature-Induced Desorption of Methyl *tert*-Butyl Ether Confined on ZSM-5: An In Situ Synchrotron XRD Powder Diffraction Study

Elisa Rodeghero [1,*], Luisa Pasti [2], Elena Sarti [2], Giuseppe Cruciani [1], Roberto Bagatin [3] and Annalisa Martucci [1]

[1] Department of Physics and Earth Sciences, University of Ferrara, Via Saragat 1, 44122 Ferrara (FE), Italy; giuseppe.cruciani@unife.it (G.C.); annalisa.martucci@unife.it (A.M.)
[2] Department of Chemical and Pharmaceutical Sciences, University of Ferrara, Via Fossato di Mortara 17, 44121 Ferrara (FE), Italy; luisa.pasti@unife.it (L.P.); elena.sarti@unife.it (E.S.)
[3] Research Centre for Non-Conventional Energy–Istituto Eni Donegani Environmental Technologies, Via Felice Maritano, 26, 20097 San Donato Milanese (MI), Italy; roberto.bagatin@eni.com
*   Correspondence: elisa.rodeghero@unife.it or rdglse@unife.it; Tel.: +39-532-974730

Academic Editor: Peng Yuan
Received: 29 December 2016; Accepted: 16 February 2017; Published: 28 February 2017

Abstract: The temperature-induced desorption of methyl *tert*-butyl ether (MTBE) from aqueous solutions onto hydrophobic ZSM-5 was studied by in situ synchrotron powder diffraction and chromatographic techniques. This kind of information is crucial for designing and optimizing the regeneration treatment of such zeolite. The evolution of the structural features monitored by full profile Rietveld refinements revealed that a monoclinic ($P2_1/n$) to orthorhombic ($Pnma$) phase transition occurred at about 100 °C. The MTBE desorption process caused a remarkable change in the unit-cell parameters. Complete MTBE desorption was achieved upon heating at about 250 °C. Rietveld analysis demonstrated that the desorption process occurred without any significant zeolite crystallinity loss, but with slight deformations in the channel apertures.

Keywords: MTBE; ZSM-5; desorption; in situ synchrotron powder diffraction

1. Introduction

The removal of methyl-*tert*-butyl-ether (MTBE, $C_5H_{12}O$) from surface waters, groundwater and urban storm water is an important goal in water treatment technology [1,2] due to the widespread occurrence of MTBE, combined with possible human carcinogenic effects [3]. This chemical is one of the main constituents of petroleum fuel and is characterized by small molecular size, high aqueous solubility (43,000–54,300 mg·L^{-1}), low Henry's law constant (0.023–0.12; dimensionless), low vapour pressure (43 mg·L^{-1} and 249 mmHg at 25 °C), and high resistance to biodegradation [4,5]. The US Environmental Protection Agency (EPA) estimated that MTBE concentration in drinking water should not exceed 20 µg·L^{-1} in terms of odour and 40 µg·L^{-1} in terms of taste [6]. Currently, water treatment technologies, such as air stripping, aerobic biodegradation, filtration, chemical oxidation reactions and membrane technology involve high operation costs and could produce toxic secondary pollutants in the environment [1,7].

Recently, adsorption on hydrophobic zeolites has received the greatest interest in water treatment technology due to their organic contaminant selectivity, thermal and chemical stability, strong mechanical properties, rapid kinetics and absence of salt and humic substance interference [8–20]. In order for the adsorption process to be cost effective, the progressive deactivation of saturated sorbents has become an essential task [20]. Thermal treatment is the most common regeneration technique, where organic host molecules are decomposed and/or oxidized at high temperature. Zeolites showed an excellent stability during the heating process and their behavior can be affected by several factors such as chemical composition, framework order-disorder and topology, nature and amount of extra-framework species, synthesis conditions, structure directing agents [21–28]. Consequently there is a strong interest in understanding the mechanisms behind the thermal regenerative solution which makes zeolites regenerable materials that are efficiently reusable in the contaminants adsorption process.

Several recent studies have demonstrated that zeolites exhibit considerably large MTBE adsorption uptakes [29–35] and that ZSM-5 was a cost-competitive adsorbent when considering both life time and usage rate of the adsorbent material [14,31]. Zeolite Socony Mobil-5 (ZSM-5, MFI-type framework topology [36]) is a medium pore material whose framework is characterized by two channel systems: sinusoidal 10-membered rings (10-MR) channels (ZZ) (sinusoidal ring A and sinusoidal ring B) along the (100) direction, interconnected with 10-MR straight channels (SC) (SC ring A and SC ring B) parallel to the (010) direction. Another tortuous pore path runs parallel to the (001) direction. The adsorption of MTBE onto high silica ZSM-5 zeolite (SiO_2/Al_2O_3 = 200) was investigated by Martucci et al. [31] by batch adsorption and X-ray powder diffraction (XRPD) analyses from aqueous solution, and by infrared spectroscopy from the gas phase in the presence of water. MTBE exhibits a type-I isotherm, thus indicating a different interaction mechanism. XRPD and infrared (IR) spectroscopy reveal the occurrence clustering of water and MTBE during adsorption from both the liquid and gas phase. These H-bonded oligomers interact with zeolite thus leading to framework flexibility for MFI-type zeolites. Rapid kinetics combined with good adsorption capacity suggest that this microporous material can be used to efficiently remove this emerging organic contaminant from water.

One of the main targets of the present work is to continuously monitor the thermal MTBE decomposition process, as well as the structural modifications on ZSM-5 upon temperature-programmed desorption treatment. This in situ synchrotron XRD powder diffraction study was used as a key to understand the features of both adsorption and desorption processes, thus helping in the design of water treatment appliances based on microporous materials.

2. Materials and Methods

2.1. Chemicals

Methyl *tert*-butyl ether (99% purity) and sodium chloride was obtained from Sigma-Aldrich (St. Louis, MO, USA). The concentration of contaminant in the aqueous solution was determined by Headspace Gas Chromatography coupled to Mass Spectrometry (HS-GC-MS). The ZSM-5 sample used in this work was a hydrophobic zeolite (code CBV 28014) provided by Zeolyst International (Conshohocken, PA, USA) in its ammonium form and used as received (SiO_2/Al_2O_3 molar ratio = 280, Na_2O < 0.05 wt % and surface area = 400 $m^2 \cdot g^{-1}$).

2.2. Experimental

The saturation capacity was determined using the batch method. Batch experiments were carried out in triplicate in 20 mL crimp top reaction glass flasks sealed with polytetrafluoroethylene (PTFE) septa (Supelco, Bellefonte, PA, USA). The flasks were filled in order to have the minimum headspace and a solid/solution ratio of 1:2 ($mg \cdot mL^{-1}$) was employed. After equilibration, for 24 h at a temperature of 25.3 ± 0.5 °C under stirring, the solids were separated from the aqueous solution by centrifugation (10,000 rpm for 30 min) and analysed by HS-SPME-GC. More details are reported in Martucci et al. [31]. The MTBE adsorbed quantities (q) and equilibrium concentrations (Ce), were determined in solution before and after equilibration with the zeolite by HS-SPME-GC.

2.3. Instrumentation

The analysis was carried out using an Agilent GC-MS system (Santa Clara, CA, USA) consisting of a GC 6850 Series II Network coupled to a Pal G6500-CTC injector and a Mass Selective Detector 5973 Network. The injected solutions consist of 100 mL of sample solutions, diluted in 10 mL of an aqueous solution saturated with NaCl, containing 10 mL of 500 mg·L^{-1} of fluorobenzene in methanol as the internal standard. HS autosampler injector conditions are as follows: incubation oven temperature 80 °C, incubation time 50 min, headspace syringe temperature 85 °C, agitation speed 250 rpm, agitation on time 30 s, agitation off time 5 s, injection volume 500 mL. In situ high-temperature X-ray diffraction data were collected at the high-resolution powder diffraction beamline ID31 (European Synchrotron Radiation Facility, ESRF, Grenoble, France). Once diffracted, the incident X-ray (λ = 0.400031 Å) was directed through nine Si 111 analyzer crystals and then collected in parallel by means of nine detectors. A subsequent data-reduction was performed to produce the equivalent step scan. X-ray diffraction patterns were recorded from room temperature to 600 °C in air (heating rate of 0.083 °C·s^{-1}), in the 0.5–19.5 2θ range. The General Structure Analysis System (GSAS) [37] package with the Experiment Graphical User Interfaces (EXPGUI) graphical interface [38] was used for Rietveld structure refinements starting from the framework fractional atomic coordinates reported by Martucci et al. [31]. The typical Rietveld fits from the temperature series and a table of the R*wp* at all temperatures are reported as Supplementary Materials (Figures S1–S17). Atomic coordinates at 30, 100 and 400 °C are also reported as Supplementary Materials (Tables S1–S3).

3. Results and Discussion

3.1. Adsorption from Aqueous Solutions

MTBE adsorption isotherms on ZSM-5 were obtained at 25 °C by Martucci et al. [31] and follow a Langmuir model. This indicates that the interactions of MTBE with the zeolite framework are energetically similar to each other. Starting from the isotherm data, saturated samples were prepared by putting into contact a given amount of zeolite with an aqueous solution of MTBE having a concentration of 200 mg·L^{-1}. The adsorbed quantity q (mg·g^{-1}) was calculated as follows:

$$q = \frac{(C_0 - C_e)V}{m} \tag{1}$$

where C_0 is the initial concentration in solution (mg·L^{-1}), C_e is the concentration at equilibrium (mg·L^{-1}), V is the solution volume (L) and m is the mass of sorbent (g). Four different samples were prepared, the average saturation capacity and standard deviation were 95 ± 7 mg·g^{-1}, confirming the saturation capacity previously found in Martucci et al. [31]. The saturated samples were homogenized and employed for the structural analysis.

After thermal desorption, the three samples of regenerated zeolite were saturated with MTBE in the same conditions. The average saturation capacity was 92 ± 8 mg·g^{-1}. Therefore, the regenerated material shows an adsorption capacity that does not significantly differ from that of the as received materials, confirming the possibility to reuse the adsorbent material.

3.2. Structural Analyses

According to Martucci et al. [31], MTBE molecules are hosted in two crystallographically independent sites: MTBE1 near the intersections of sinusoidal and straight 10 MR channels (C1a, C2a, C3a, C4a, C5a, O1a sites in Table S1); and MTBE2 in the sinusoidal 10 MR channel (C1b, C2b, C3b, C4b, C5b, O1b sites in Table S1), respectively. Figure 1 shows the high-loaded structure of ZSM-5-MTBE, with the guest molecules located in both channels. On the whole, eight MTBE molecules (corresponding to ~11% zeolite dry weight (dw)) and about two water molecules (corresponding to about 0.5% zeolite dw) were detected. Rietveld structure refinement confirmed the occurrence of

MTBE–water complexes interacting with the framework, stabilizing the guest structure within the zeolite host framework. The presence of organic compound–water molecule oligomers has also been recently reported in mordenite [39], ferrierite [29] as well as in the same Y and ZSM-5 zeolite after 1,2-dichloroethane (DCE) adsorption from aqueous solutions [18].

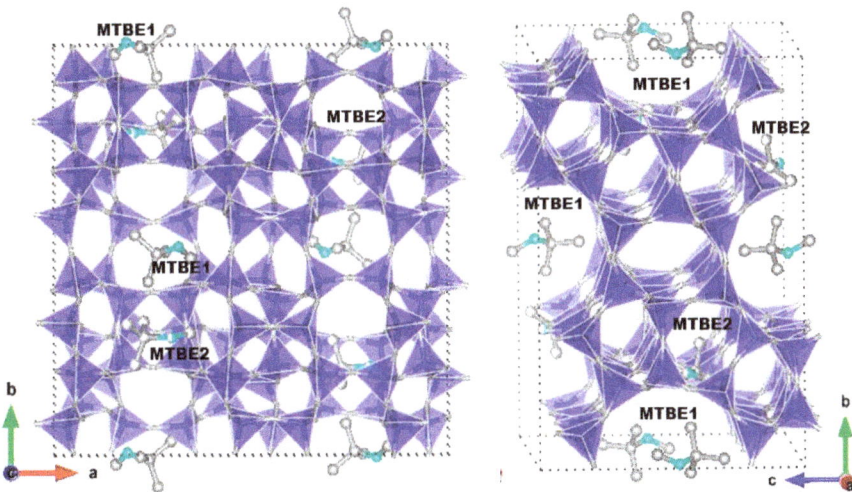

Figure 1. High-loaded structure of ZSM-5-methyl-*tert*-buthyl-ether (MTBE) along *a* and *c* directions, respectively. Water molecules are represented as light blue spheres.

The automatic indexing of the peaks, carried out by the High Score Plus v. 3.0 software [40], revealed the gradual overlapping of groups of peaks (i.e., $131 + 13 - 1$ and $311 + 31 - 1$ in the first angular range, and $133 + 13 - 3$ and $313 + 31 - 3$ in the second range) attesting the monoclinic to orthorhombic phase transition, with a Tc close to $100 \pm 5\,°C$. Figure 2 shows the evolution of the investigated ZSM-5 sample close to the expected transition temperature, Tc, in the 3.70–4.30 and 5.60–6.50 2θ range. Therefore, recent works [41–43] reported this phenomenon both in the unloaded ZSM-5 as well as in the same samples after organics adsorption. The evolution of refined unit cell parameters as a function of temperature is illustrated in Figure 2.

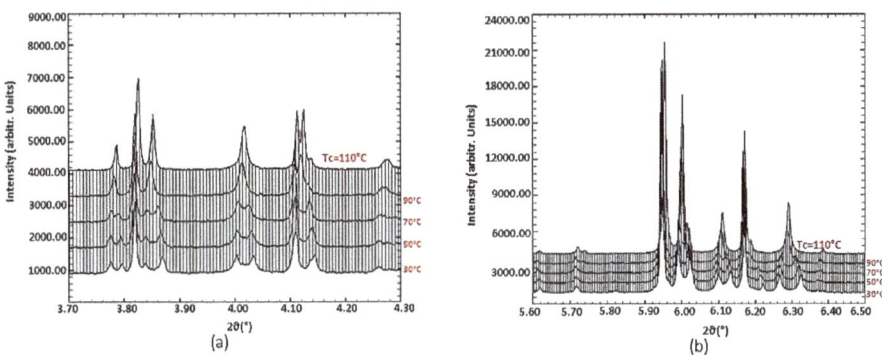

Figure 2. Evolution of the investigated ZSM-5-MTBE close to the expected transition temperature (30, 50, 70, 90 and 110 °C, respectively) in the 3.70–4.30 (**a**) and 5.60–6.50 (**b**) 2θ (°) range.

The general trend shows an initial increase of all the lattice parameters except for the b parameter (Figure 3 and Table S1). In particular, the unit cell volume increases until about 125 ± 5 °C, then it remains about constant in the range between 125 and 200 ± 5 °C and starts to decrease after 200 ± 5 °C. Similar behaviour was observed during the desorption process of 1,2-dichloroethane [44] and toluene [45].

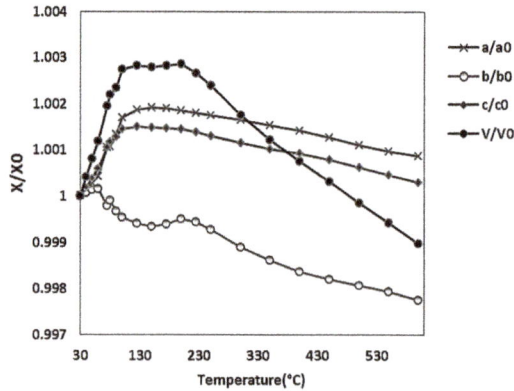

Figure 3. Temperature evolution of ZSM-5-MTBE unit cell parameters during in situ thermal organic burning. All values are normalized compared to those refined at room temperature.

According to Pasti et al., Martucci et al. [44] and Rodeghero et al. [45] this behavior can be explained by the relaxation of the interactions between the oligomers and the framework oxygens. The refined distances clearly indicate that during the heating, this attractive force exerting a negative pressure is released and the framework is free to relax and expand. The process is also correlated to the MTBE desorption process which starts at 100 °C and is complete at ~300 °C. Figure 4 shows both the evolution of MTBE molecules per unit cell (p.u.c.) and the unit cell volume as a function of temperature.

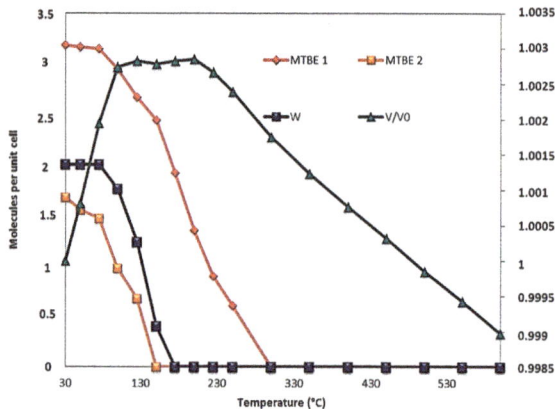

Figure 4. Evolution of MTBE and water (W) molecules and unit cell volume as a function of temperature.

After the full desorption of organics, a contraction of the unit cell volume is observed until 600 °C. This phenomenon is explained as a thermal negative expansion (NTE), already observed not only in MFI-type materials [41,45–48], but also in other microporous materials [22–27,48–51], thus attesting the relaxation of framework distortions induced by host molecules which diffuse through the zeolite channels during the heating process. The desorption process occurred without any significant zeolite crystal. Above 300 °C, the channels ellipticity decreases and wide-open apertures regularize. This process is accompanied by variations in the opening of the zeolite framework pore system and consequently, in the Crystallographic Free Areas (C.F.A.) (Figure 5).

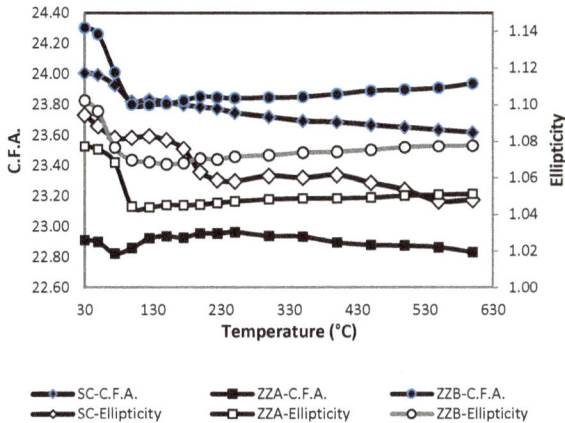

Figure 5. Crystallographic Free Areas (C.F.A.) of the 10-ring channels and ellipticity of the 10-ring channels as a function of temperature. ZZ stands for zigzag or sinusoidal channel; SC stands for straight channel.

4. Conclusions

This study reports experimental results concerning the desorption of methyl-*tert*-butyl-ether from the pores of a silica-rich zeolite ZSM-5. The temperature-induced desorption of this oxygenated compound from aqueous solutions onto hydrophobic ZSM-5 was studied by combining in situ synchrotron powder diffraction and chromatographic techniques. The evolution of the structural features monitored by full profile Rietveld refinements revealed that a monoclinic ($P2_1/n$) to orthorhombic (*Pnma*) phase transition occurred at about 100 °C. Complete MTBE desorption was achieved upon heating at about 250 °C. Notwithstanding the change in the unit-cell parameters, Rietveld refinement demonstrated that the desorption process occurred without any significant zeolite crystallinity loss, but with slight deformations in the channel apertures. After the full organics desorption, a contraction of the unit cell volume is observed, thus indicating negative thermal expansion (NTE) for this material. On the basis of all these results, ZSM-5 represents a promising adsorbent medium to remove MTBE contaminant from water.

Supplementary Materials: The following are available online at www.mdpi.com/2075-163X/7/3/34/s1, Figures S1–S17: Observed (dotted upper line), calculated (solid upper line), and difference (solid lower line) powder diffraction patterns of ZSM-5 at 30, 50, 75, 90, 125, 150, 175, 200, 225, 250, 300, 350, 400, 450, 500, 550 and 600 °C. Crystallographic data from the Rietveld refinement are also reported; Tables S1–S3: Fractional atomic coordinates of ZSM-5 loaded with MTBE at Room Temperature (30 °C), 100 and 400 °C.

Acknowledgments: Research co-funded by the Research Centre for Unconventional Energies, Istituto ENI G. Donegani-Environmental Technologies (San Donato Milanese (MI), Italy). Elisa Rodeghero, Annalisa Martucci and Giuseppe Cruciani also acknowledge MIUR for funding support within PRIN2010 programme (prot. 2010EARRRZ_009). We would like to acknowledge the also the European Synchrotron Radiation Facility (ESRF, Grenoble) for providing beam time (proposal CH-3510—In situ XRD study of structural modifications and desorption kinetics of zeolites used for removal of non-polar organic compounds from contaminated water).

Author Contributions: The manuscript was written with contributions from all authors. All authors have given approval to the final version of the manuscript. Elisa Rodeghero wrote the paper and performed the X-ray experiments with Giuseppe Cruciani; Luisa Pasti, Roberto Bagatin and Annalisa Martucci conceived and designed the experiments and analyzed the data.

Conflicts of Interest: The authors declare no conflict of interest.

References

1. Chong, M.N.; Jin, B.; Chow, C.W.; Saint, C. Recent developments in photocatalytic water treatment technology: A review. *Water Res.* **2010**, *44*, 2997–3027. [CrossRef] [PubMed]
2. Johnson, R.; Pankow, J.; Bender, D.; Price, C.; Zogorski, J. MTBE—To What Extent Will Past Releases Contaminate Community Water Supply Wells? *Environ. Sci. Technol.* **2000**, *34*, 210A–217A. [CrossRef] [PubMed]
3. Amberg, A.; Rosner, E.; Dekant, W. Toxicokinetics of methyl *tert*-butyl ether and its metabolites in humans after oral exposure. *Toxicol. Sci.* **2001**, *61*, 62–67. [CrossRef] [PubMed]
4. Mackay, D.; Shiu, W.Y.; Ma, K.C.; Lee, S.C. *Illustrated Handbook of Physical–Chemical Properties and Environmental Fate for Organic Chemicals—Volatile Organic Chemicals*; Taylor & Francis Group, LLC: Boca Raton, FL, USA, 2006; Volume 3, pp. 1–925.
5. Squillace, P.; Pankow, J.; Kortes, N.; Zogorski, J. Review of the environmental behavior and fate of methyl *tert*-butyl ether. *Environ. Toxicol. Chem.* **2009**, *16*, 1836–1844. [CrossRef]
6. United States Environmental Protection Agency. *Drinking Water Advisory: Consumer Acceptability Advice and Health Effects Analysis on Methyl Tertiary-Butyl Ether (MTBE)*; EPA-822-F-97-009; US Environmental Protection Agency: Washington, DC, USA, 1997; pp. 11–13.
7. Gaya, U.I.; Abdullah, A.H. Heterogeneous photocatalytic degradation of organic contaminants over titanium dioxide: A review of fundamentals, progress and problems. *J. Photochem. Photobiol. C Photochem. Rev.* **2008**, *9*, 1–12. [CrossRef]
8. Costa, A.A.; Wilson, W.B.; Wang, H.; Campiglia, A.D.; Dias, J.A.; Dias, S.C.L. Comparison of BEA, USY and ZSM-5 for the quantitative extraction of polycyclic aromatic hydrocarbons from water samples. *Microporous Mesoporous Mater.* **2012**, *149*, 186–192. [CrossRef]
9. Milestone, N.B.; Bibby, D.M. Concentration of alcohols by adsorption on silicalite. *J. Chem. Technol. Biotechnol.* **1981**, *31*, 732–736. [CrossRef]
10. Grose, R.W.; Flanigen, E.M. Novel Zeolite Compositions and Processes for Preparing and Using Same. U.S. Patent 4,257,885, 24 March 1981.
11. Yazaydin, A.O.; Thompson, R.W. Molecular simulation of the adsorption of MTBE in silicalite, mordenite, and zeolite beta. *J. Phys. Chem. B* **2006**, *110*, 14458–14462. [CrossRef] [PubMed]
12. Abu-Lail, L.; Bergendahl, J.A.; Thompson, R.W. Adsorption of methyl tertiary butyl ether on granular zeolites: Batch and column studies. *J. Hazard. Mater.* **2010**, *178*, 363–369. [CrossRef] [PubMed]
13. Anderson, M.A. Removal of MTBE and other organic contaminants from water by sorption to high silica zeolites. *Environ. Sci. Technol.* **2000**, *34*, 725–727. [CrossRef]
14. Rossner, A.; Knappe, D.R. MTBE adsorption on alternative adsorbents and packed bed adsorber performance. *Water Res.* **2008**, *42*, 2287–2299. [CrossRef] [PubMed]
15. Martucci, A.; Braschi, I.; Marchese, L.; Quartieri, S. Recent advances in clean-up strategies of waters polluted with sulfonamide antibiotics: A review of sorbents and related properties. *Mineral. Mag.* **2014**, *78*, 1115–1140. [CrossRef]
16. Braschi, I.; Blasioli, S.; Gigli, L.; Gessa, C.E.; Alberti, A.; Martucci, A. Removal of sulfonamide antibiotics from water: Evidence of adsorption into an organophilic zeolite Y by its structural modifications. *J. Hazard. Mater.* **2010**, *17*, 218–225. [CrossRef] [PubMed]
17. Martucci, A.; Pasti, L.; Marchetti, N.; Cavazzini, A.; Dondi, F.; Alberti, A. Adsorption of pharmaceuticals from aqueous solutions on synthetic zeolites. *Microporous Mesoporous Mater.* **2012**, *148*, 174–183. [CrossRef]
18. Pasti, L.; Martucci, A.; Nassi, M.; Cavazzini, A.; Alberti, A.; Bagatin, R. The role of water in DCE adsorption from aqueous solutions onto hydrophobic zeolites. *Microporous Mesoporous Mater.* **2012**, *160*, 182–193. [CrossRef]

19. Pasti, L.; Sarti, E.; Cavazzini, A.; Marchetti, N.; Dondi, F.; Martucci, A. Factors affecting drug adsorption on beta zeolites. *J. Sep. Sci.* **2013**, *36*, 1604–1611. [CrossRef] [PubMed]

20. Braschi, I.; Martucci, A.; Blasioli, S.; Mzini, L.L.; Ciavatta, C.; Cossi, M. Effect of humic monomers on the adsorption of sulfamethoxazole sulfonamide antibiotic into a high silica zeolite Y: An interdisciplinary study. *Chemosphere* **2016**, *155*, 444–452. [CrossRef] [PubMed]

21. Leardini, L.; Martucci, A.; Braschi, I.; Blasioli, S.; Quartieri, S. Regeneration of high-silica zeolites after sulfamethoxazole antibiotic adsorption: A combined in situ high-temperature synchrotron X-ray powder diffraction and thermal degradation study. *Mineral. Mag.* **2014**, *78*, 1141–1160. [CrossRef]

22. Grima, J.N.; Zammit, V.; Gatt, R.M. Negative Thermal Expansion. *Xjenza* **2006**, *11*, 17–29.

23. Bull, I.; Lightfoot, P.; Villaescusa, L.A.; Bull, L.M. An X-ray Diffraction and MAS NMR Study of the Thermal Expansion Properties of Calcined Siliceous Ferrierite. *J. Am. Chem. Soc.* **2003**, *125*, 4342–4349. [CrossRef] [PubMed]

24. Woodcock, D.A.; Lightfoot, P.; Wright, P.A.; Villaescusa, L.A.; Dìaz-Cabañasb, M.J.; Camblorb, M.A. Strong negative thermal expansion in the siliceous zeolites ITQ-1,ITQ-3 and SSZ-23. *J. Mater. Chem.* **1999**, *9*, 349–351. [CrossRef]

25. Alberti, A.; Martucci, A. Phase transformations and structural modifications induced by heating in microporous materials. *Stud. Surf. Sci. Catal.* **2005**, *155*, 19–43.

26. Alberti, A.; Martucci, A. Reconstructive phase transitions in microporous materials: Rules and factors affecting them. *Microporous Mesoporous Mater.* **2011**, *141*, 192–198. [CrossRef]

27. Cruciani, G. Zeolites upon heating: Factors governing their thermal stability and structural changes. *J. Phys. Chem. Solids* **2006**, *67*, 1973–1994. [CrossRef]

28. Leardini, L.; Martucci, A.; Alberti, A.; Cruciani, G. Template burning effects on stability and boron coordination in boron levyne studied by in situ time resolved synchrotron powder diffraction. *Microporous Mesoporous Mater.* **2013**, *167*, 117–126. [CrossRef]

29. Martucci, A.; Leardini, L.; Nassi, M.; Sarti, E.; Bagatin, R.; Pasti, L. Removal of emerging organic contaminants from aqueous systems: Adsorption and location of methyl-tertiary-butylether on synthetic ferrierite. *Mineral. Mag.* **2014**, *78*, 1161–1175. [CrossRef]

30. Arletti, R.; Martucci, A.; Alberti, A.; Pasti, L.; Nassi, M.; Bagatin, R. Location of MTBE and toluene in the channel system of the zeolite mordenite: Adsorption and host–guest interactions. *J. Solid State Chem.* **2012**, *194*, 135–142. [CrossRef]

31. Martucci, A.; Braschi, I.; Bisio, C.; Sarti, E.; Rodeghero, E.; Bagatin, R.; Pasti, L. Influence of water on the retention of methyl tertiary-butyl ether by high silica ZSM-5 and Y zeolites: A multidisciplinary study on the adsorption from liquid and gas phase. *RSC Adv.* **2015**, *5*, 86997–87006. [CrossRef]

32. Knappe, D.R.U.; Campos, A.A.R. Effectiveness of high-silica zeolites for the adsorption of methyl tertiary-butyl ether from natural water. *Water Sci. Technol. Water Supply* **2005**, *5*, 83–91.

33. Centi, G.; Grande, A.; Perathoner, S. Catalytic conversion of MTBE to biodegradable chemicals in contaminated water. *Catal. Today* **2002**, *75*, 69–76. [CrossRef]

34. Sacchetto, V.; Gatti, G.; Paul, G.; Braschi, I.; Berlier, G.; Cossi, M.; Bisio, C. The interactions of methyl *tert*-butyl ether on high silica zeolites: A combined experimental and computational study. *Phys. Chem. Chem. Phys.* **2013**, *15*, 13275–13287. [CrossRef] [PubMed]

35. Li, S.; Tuan, V.A.; Noble, R.D.; Falconer, J.L. MTBE adsorption on all-silica β zeolite. *Environ. Sci. Technol.* **2003**, *37*, 4007–4010. [CrossRef] [PubMed]

36. Baerlocher, C.; Meir, W.M.; Olson, O.H. *Atlas of Zeolite Framework Types*, 5th ed.; Elsevier Science: New York, NY, USA, 2001.

37. Larson, A.C.; von Dreele, R.B. *GSAS, General Structure Analysis System*; LANSCE, MS-H805; Los Alamos National Laboratory: Los Alamos, NM, USA, 1994.

38. Toby, B.H. EXPGUI, a graphical user interface for GSAS. *J. Appl. Crystallogr.* **2001**, *34*, 210–213. [CrossRef]

39. Martucci, A.; Pasti, L.; Nassi, M.; Alberti, A.; Arletti, R.; Bagatin, R.; Sticca, R. Adsorption mechanism of 1,2-dichloroethane into an organophilic zeolite mordenite: A combined diffractometric and gas chromatographic study. *Microporous Mesoporous Mater.* **2012**, *151*, 358–367. [CrossRef]

40. Degen, T.; Sadki, M.; Bron, E.; König, U.; Nénert, G. The HighScore suite. *Powder Diffr.* **2014**, *29*, S13–S18. [CrossRef]

41. Ardit, M.; Martucci, A.; Cruciani, G. Monoclinic–orthorhombic phase transition in ZSM-5 zeolite: Spontaneous strain variation and thermodynamic properties. *J. Phys. Chem. C* **2015**, *119*, 7351–7359. [CrossRef]

42. Martucci, A.; Rodeghero, E.; Pasti, L.; Bosi, V.; Cruciani, G. Adsorption of 1,2-dichloroethane on ZSM-5 and desorption dynamics by in situ synchrotron powder X-ray diffraction. *Microporous Mesoporous Mater.* **2015**, *215*, 175–182. [CrossRef]

43. Rodeghero, E.; Martucci, A.; Cruciani, G.; Bagatin, R.; Sarti, E.; Bosi, V.; Pasti, L. Kinetics and dynamic behaviour of toluene desorption from ZSM-5 using in situ high-temperature synchrotron powder X-ray diffractionand chromatographic techniques. *Catal. Today* **2016**, *227*, 118–125. [CrossRef]

44. Bhange, D.S.; Ramaswamy, V. High temperature thermal expansion behavior of silicalite-1 molecular sieve: In Situ HTXRD study. *Microporous Mesoporous Mater.* **2007**, *103*, 235–242. [CrossRef]

45. Bhange, D.S.; Ramaswamy, V. Enhanced negative thermal expansion in MFI molecular sieves by varying framework composition. *Microporous Mesoporous Mater.* **2010**, *130*, 322–326. [CrossRef]

46. Villaescusa, L.A.; Lightfoot, P.; Teat, S.J.; Morris, R.E. Variable-Temperature Microcrystal X-ray Diffraction Studies of Negative Thermal Expansion in the Pure Silica Zeolite IFR. *J. Am. Chem. Soc.* **2001**, *123*, 5453–5459. [CrossRef] [PubMed]

47. Milanesio, M.; Artioli, G.; Gualtieri, A.F.; Palin, L.; Lamberti, C. Template burning inside TS-1 and Fe-MFI molecular sieves: An in situ XRPD study. *J. Am. Chem. Soc.* **2003**, *125*, 14549–14558. [CrossRef] [PubMed]

48. Martucci, A.; de Lourdes Guzman-Castillo, M.; Di Renzo, F.; Fajula, F.; Alberti, A. Reversible channel deformation of zeolite omega during template degradation highlighted by in situ time-resolved synchrotron powder diffraction. *Microporous Mesoporous Mater.* **2007**, *104*, 257–268. [CrossRef]

49. Leardini, L.; Martucci, A.; Cruciani, G. The unusual thermal behaviour of boron-ZSM-5 probed by "in situ" time-resolved synchrotron powder diffraction. *Microporous Mesoporous Mater.* **2013**, *173*, 6–14. [CrossRef]

50. Leardini, L.; Martucci, A.; Cruciani, G. The unusual thermal expansion of pure silica sodalite probed by in situ time-resolved synchrotron powder diffraction. *Microporous Mesoporous Mater.* **2012**, *151*, 163–171. [CrossRef]

51. Leardini, L.; Quartieri, S.; Vezzalini, G.; Arletti, R. Thermal behaviour of siliceous faujasite: Further structural interpretation of negative thermal expansion. *Microporous Mesoporous Mater.* **2015**, *202*, 226–233. [CrossRef]

minerals

MDPI

Review

Application of Mineral Sorbents for Removal of Petroleum Substances: A Review

Lidia Bandura [1,*]**, Agnieszka Woszuk** [2]**, Dorota Kołodyńska** [3] **and Wojciech Franus** [1]

1   Department of Geotechnical Science, Faculty of Civil Engineering and Architecture,
    Lublin University of Technology, Nadbystrzycka 40, 20-618 Lublin, Poland; w.franus@pollub.pl
2   Department of Roads and Bridges, Faculty of Civil Engineering and Architecture,
    Lublin University of Technology, Nadbystrzycka 40, 20-618 Lublin, Poland; a.woszuk@pollub.pl
3   Department of Inorganic Chemistry, Faculty of Chemistry, Maria Curie-Skłodowska University,
    Maria Curie-Skłodowska Sq.2, 20-031 Lublin, Poland; d.kolodynska@poczta.umcs.lublin.pl
*   Correspondence: l.bandura@pollub.pl; Tel.: +48-509-381-660

Academic Editor: Annalisa Martucci
Received: 15 December 2016; Accepted: 4 March 2017; Published: 8 March 2017

Abstract: Environmental pollution with petroleum products has become a major problem worldwide, and is a consequence of industrial growth. The development of sustainable methods for the removal of petroleum substances and their derivatives from aquatic and terrestrial environments and from air has therefore become extremely important today. Advanced technologies and materials dedicated to this purpose are relatively expensive; sorption methods involving mineral sorbents are therefore popular and are widely described in the scientific literature. Mineral materials are easily available, low-cost, universal adsorbents and have a number of properties that make them suitable for the removal of petroleum substances. This review describes recent works on the use of natural, synthetic and modified mineral adsorbents for the removal of petroleum substances and their derivatives from roads, water and air.

Keywords: petroleum substances; oils; BTEX; sorption; removal; mineral adsorbents

1. Introduction

In the modern context of advanced and developing industrialization, petroleum products and their derivatives constitute one of the major sources of environmental pollution. During the extraction, transport, distribution and storage of crude oil and its products, these may be released into the environment in an uncontrolled manner, causing pollution of the atmosphere, lithosphere, hydrosphere and biosphere [1–6]. Due to the extent of impact and the adverse effect of oil derivatives on both the inanimate and animate environment, and the limits imposed on their emissions and permissible concentrations in soils and water, the search for effective methods and new materials for the removal of such substances from contaminated sites is extremely important.

Environmental pollution caused by petroleum substances can be divided into four groups, according to location:

- atmospheric pollution caused by the evaporation of volatile components of petroleum products;
- pollution of soils;
- pollution of aquatic systems;
- environmental pollution caused by land-based spills of petroleum products.

For each of these groups, it is important to choose appropriate methods for the removal of undesirable organic substances in the most efficient way.

A number of mechanical, biological, chemical and adsorption methods are currently used to remove oil spills from water media and paved roads, and to remove volatile hydrocarbons (Figure 1) [7–14].

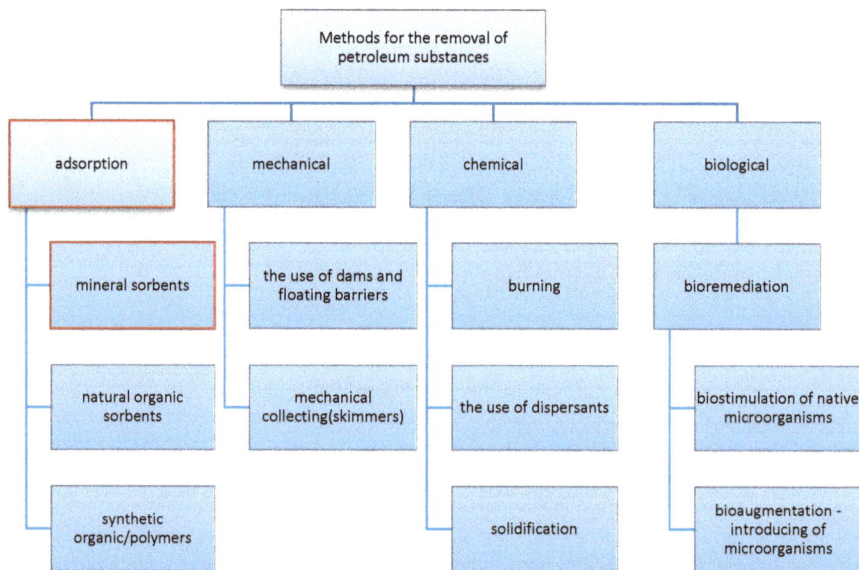

Figure 1. Methods used for the petroleum substance removal.

Solutions applied to volatile organic compounds are primarily aimed at monitoring and reducing their emissions using destructive methods and recovery techniques [15]. The most popular of these are adsorption techniques involving various kinds of adsorbents.

Usually, during rescue operations on waters, more than one method is applied. Comprehensive rescue operations are aimed at efficient removal of oil spills from the water surface and the coastal zones and at preventing further migration of these contaminants. The different cleanup techniques are discussed as follows [16]:

In situ burning is performed as soon as possible after the spill, before the oil stain will begin to evaporate volatile components and disperse. Usually, burning is used at large water surfaces (seas and oceans) but also on land. To perform the burning process effectively the oil spill should be fresh, contain volatile components, and the thickness of the oil layer should be at least 2–3 mm. In addition, the water surface cannot be rough, and the location of burning should be safely away from the residential areas, flammable objects, the coasts, living organisms habitats, etc. Such conditions occur in Arctic areas, since low temperatures limit the movement of water, the dispersion of stains, and evaporation of volatile components [17]. The main advantages of in situ burning are: low costs, high removal efficiency in a relatively short time (100–300 tons per hour). A major drawbacks are the atmospheric emissions of toxic combustion products (CO, SO_2, polycyclic aromatic hydrocarbons—PAHs) and the threat of the environment.

Dispersive method involves the removal of residual oil film with the use of surface-active agents (surfactants). Due to an amphiphilic character, they reduce the interfacial tension between oil and water causing dispersion and oil dilution. Thus, further process of natural biodegradation of petroleum compounds is facilitated and the subsequent removal by mechanical, biological or sorption means is easier and more effective [18,19]. Due to toxicity the usage of dispersants in some countries is limited (UK, USA). Dispersion method is efficient in the removal of large scale oil spills (it is possible to spray dispersants from a helicopter). They are not recommended in the case of heavy, high viscosity oils or in cold, calm water. Surfactants are also used in cleaning soils from petroleum derivatives [20].

Solidifiers are generally dry, granular, hydrophobic organic polymers that react with oil to form a monolithic solid that floats on water [21,22]. The formed solid can be easily removed from the water surface. The effectiveness of solidification depends on the ambient temperature, the type and chemical composition of the oil (solidifier can bind only one type of petroleum substance) which may limit their application. Furthermore, it is recommended to use large amounts of the material (16%–200% by weight) in relation to the weight of oil spill [23]. Few recent studies in this area indicate that solidification is possible only to local impurities of small range. Mixing solidifiers with the oil is difficult, solidifiers may exhibit toxic properties, there is also a problem with formed solid waste disposal. Sustainable methods for recycling of spent solidifiers have not yet been developed [24].

Mechanical methods belong to the more expensive solutions because they often require the use of specialized equipment. However, they are far less harmful to the environment than chemical methods. These include the application of booms, skimmers, floating dams and barriers in order to limit further spread of oil and to collect it mechanically from the water surface [10,25]. Mechanical methods are only effective in case of stagnant water and calm seas, with little wind. The structural designs of skimmers have a great impact on their efficiency. Skimmers allow for the recovery of oil but there exists a risk of oil leakage.

Biological methods are usually applied at the final stage of the purification operations in the aquatic environments or in case of increased acceptable level of pollution in the waters. Their aim is to purify water using microorganisms that are able to decompose petroleum substances and restore the biological balance of aquatic ecosystems. Biological methods are also dedicated to soils polluted with oil derivatives. To this end, strains of native microorganisms are often augmented which seems to be the most efficient in hydrocarbons decomposition in particular localization. For an extensive overview of the topic, readers are sent to the papers: [26–31].

The described methods were applied during the main environmental disasters involving oil spillages from: Liberian tanker at Chedabucto Bay (1971), Exxon Valdez tanker in Prince Williams Sound Alaska (1989), Deepwater Horizon/BP oil rig explosion in Gulf of Mexico (2010). In addition to them, various types of adsorbents were also used.

Sorption methods involve the use of adsorbents which are porous solids with developed specific surface area, capable of binding molecules (from the liquid or gaseous phase) on its surface. Thus, these materials have a wide range of applications, especially in various types of purification technologies. Adsorbents are used for petroleum derivatives removal from coastal areas, waters, paved roads, exhaust gases and vapors. Generally, sorption methods are considered as most effective, inexpensive, available, resistant for atmospheric conditions, easy and safe. In chemical rescue, adsorbents are commonly used to immobilize and remove spills of hazardous liquids from water or paved surfaces. They are also used in situations where it is necessary to remove residual contaminants remaining in the environment after mechanical oil collection, as well as a barrier to prevent the further spread of dangerous liquids (a protective embankment). In most cases, they can be recycled or adsorbed substance can be recovered from them.

The appropriate adsorbent should be chosen depending on the location of the accident and the type and quantity of spilled petroleum substance. In the selection of adsorbent, the following criteria should be taken into account: the sorption capacity of the adsorbent, its ability to immobilize oil, its buoyancy, efficiency, availability and biodegradability, possibility of recycling and/or reuse, environmental impact, price, non-flammability and resistance to chemicals and environmental conditions [8].

Al-Majed et al. [16] have proposed a sustainable pathway for methods of oil spills control. It involves eliminating or reducing chemical methods which are dangerous for the environment through replacing them with most eco-friendly techniques such as mechanical methods and adsorbents.

The aim of this paper is to present the potential application of natural minerals and synthetic zeolites as the adsorbents for the removal of petroleum substances. Crude oil and petroleum products belong to hazardous environmental pollutants. Because of the extend of impact and adverse effect of oil derivatives on the inanimate and animate environment, it is extremely important to seek effective methods and materials for the removal of such substances from contaminated areas. What is more, proposed solutions should follow the principles of sustainable development which means that the adsorbents have to be easy-available, inexpensive, recyclable or easy to recover, and non-toxic. Mineral adsorbents meet this criteria to a large extent. Recently, numerous research articles on mineral sorbents have been published, in terms of their usage in the removal of petroleum products and the oil spills cleanup. This review summarizes the current results (from the years 2010−2016) on mineral materials and synthetic zeolites that have been used in their raw form and/or after thermal and chemical modifications in the sorption of oils and petroleum compounds from roads, waters and gases.

2. Adsorbents in the Petroleum Spills Cleanup

According to Adebajo et al. [32], adsorbents used in the removal of oil spills can be divided into three main groups, depending on the source of origin:

(1) inorganic mineral sorbents;
(2) natural organic sorbents;
(3) synthetic organic sorbents (synthetic polymers).

Mineral adsorbents represent a very large group; these are commonly used as they have a number of advantages such as non-flammability, chemical inertness, relatively low cost and easy availability. These are also known as sinking sorbents, and they are highly dense, fine-grained materials—natural or processed—used to sink floating oil. They can be considered as a group of universal adsorbents. Most mineral adsorbents are raw materials of natural origin which are used in a powder or granular form. Their particle size may range from several nm to several mm (not more than 3 mm). Mineral adsorbents are generally non-combustible and resistant to acids and bases. Usually, their sorption capacity towards petroleum derivatives is in the range 0.20–0.50 g/g, and their bulk density is 0.45–0.90 kg/dm^3 [33]. Their primary disadvantage is a risk of dust formation during application in open spaces, and respiratory protective equipment and eye protection are therefore required when using certain powder sorbents. Mineral sorbents are poured onto a (land-based) oil spill, spread onto the surface mechanically (using a brush), allowed to absorb the substance, and are then collected together with the absorbed substance and transferred for recycling. They are not generally preferred for the removal of oil spills from water surfaces because of their low buoyancy, and low oil absorbability comparable to polymers or natural organic sorbents [34]. However, their modification with organic compounds are reported as suitable for water media.

Natural organic adsorbents used in chemical rescue include peat, needle-cover, moss, dry leaves, straw, sawdust, bark and wood waste, cellulose from paper and cotton products, linen materials, cotton and hemp [33]. The literature also describes other sorbents of natural origin from agricultural and/or processing wastes, such as rice husk, various types of plant shells and plant waste, kapok and many others [34–38]. Natural organic adsorbents are considered to be effective, inexpensive (although this is not universally true), easy available and environmentally friendly. They are biodegradable and flammable, and thus are easy to utilize. However, their low bulk density and lightness may cause an impediment in open spaces, and their poor buoyancy limits their use in an aqueous environment. It is impossible to use them in the case of fire. Of the natural organic adsorbents, the most efficient are those subjected to special treatment (e.g., thermal), and these exhibit sorption capacity towards oils in the range 0.7–4.0 g/g; some papers report sorption performance even/much higher that this [32,33].

The group of synthetic polymers includes polypropylene, polyethylene, polyacrylate, polystyrene, and polyurethane, which are used to manufacture special sleeves, mats, cloths, or cushions for the sorption of hazardous liquids. Polymer adsorbents exhibit hydrophobic properties, low bulk density

(0.10–0.45 kg/dm^3), and large sorption capacity with respect to petroleum derivatives. Depending on the type of material, the sorption capacity ranging from few to several tens g/g, and some studies indicate that the capacity can exceed 100 g/g [39–41]. Due to their buoyancy and hydrophobicity they are mainly used in aqueous media and very rarely for the removal of oil spills from rigid pavements, since they are often too light and easily blown by the wind. Further disadvantages of these materials are the possibility of returning the absorbed liquid under external forces, non-biodegradability and flame retardant properties. Their utilization also involves the problem of emission of toxic compounds during combustion. Although methods which allow the recovery of the synthetic polymer sorbents after oil sorption, such as centrifugation and pressing, are known, these are very limited due to the destruction of the structure of the sorbent, oxidation or strong contamination.

2.1. Mineral Adsorbents

2.1.1. Zeolites

Zeolites are aluminosilicates of the alkaline and alkaline earth metals. The main elements of the crystalline framework of zeolites are $[SiO_4]^{4-}$ and $[AlO_4]^{3-}$ tetrahedrons connected by oxygen atoms. These tetrahedrons form three-dimensional lattice with free channels of diameter 0.3–3 nm, which gives these minerals a "molecular sieve" character and sorption properties. A negative charge is created as a consequence of the partial replacement of Si^{4+} ions by Al^{3+} ions in the zeolites' crystal lattice, which is compensated for by Ca^{2+}, Na^+ or K^+ ions localized in channels where H_2O molecules are also present. The cations are readily replaced by others from the surrounding solution, thus giving rise to the ion exchange capabilities of zeolites. In addition, surface OH groups provide these minerals with acid and sorption properties. From the perspective of their origin, zeolites can be divided into natural (created/formed as a result of geological processes occurring in nature) and synthetic. The synthetic zeolites are usually obtained from the chemical reaction between sodium silicate, Na_2SiO_3, and sodium aluminate, $NaAlO_2$, under varying conditions of temperature, pressure and reaction time. In addition to pure chemical reagents, raw materials such as fly ash, perlite, clay minerals and obsidian can be used for the synthesis of zeolites [42–46].

2.1.2. Clay Minerals

Clay minerals cover several groups of hydrous aluminium phyllosilicates. They form in sediments, soils and as the result of the diagenetic and hydrothermal alteration of rocks. Their foundation is formed from $[SiO_4]^{4-}$ tetrahedrons, connected at three corners by shared oxygen anions, thus forming the tetrahedral sheet. Divalent or trivalent metal cations (aluminium, magnesium, iron and calcium) are bonded to the tetrahedral sheet, coordinated to one hydroxyl and two oxygen anion groups and surrounded by six oxygens or hydroxyl groups, thus forming the octahedral sheet. Depending on the arrangement of the tetrahedral and octahedral sheets, phyllosilicate ratios of 1:1 (involving units of alternating tetrahedral and octahedral sheets), 2:1 (two tetrahedral and one octahedral sheet) and 2:1:1 (two 2:1 layers with one octahedral sheet between them) can be distinguished [47,48]. In the structure of clay minerals the interlayer space is occupied by hydrated cations. The most common representatives of various structures of clay minerals are kaolinite, montmorillonite and sepiolite. Clay minerals exhibit many special properties such as sorption capacity, swelling behaviour, and ion exchange capability which result from their unique structure, the presence of surface OH groups and weak electrostatic interactions between the layers/sheets and the exchangeable cations.

2.1.3. Silica Adsorbents

The group of silica adsorbents includes rocks (siliceous earths, diatomaceous earths, diatomites) and perlite. In terms of their mineral composition, mineraloids dominate this group (opal and chalcedony), with certain amounts of other minerals being present (cristobalite, quartz, clay minerals and carbonates). Opal can occur in a colloidal silica form, with a variable water content (1–21 wt %)

which is released in a continuous manner during drying of the mineral. It is an amorphous substance containing a disordered skeleton of $[SiO_4]^{4-}$ tetrahedrons with H_2O molecules located in the voids. Opal thus transforms to chalcedony, a semi-crystalline type of quartz. The sorption properties of silica adsorbents result from their significant porosity and the presence of surface hydroxyl groups. These groups are created during the natural formation of opal, and as a result of chemical reactions between the mineral surface and substances present in the environment. Diatomaceous earth and diatomites are formed from the exoskeletons of unicellular algae known as diatoms, which collected at the bottom of water bodies over many millions of years. The exoskeletons of these microorganisms form a unique structure which provides the mineral with a significant contribution of free spaces (between exoskeletons), meso- and macroporosity. Thus, diatomaceous materials are widely used as sorbents of petroleum compounds in the form of oils and vapours of organic compounds [33] including benzene, toluene, ethylbenzene and xylenes (BTEX) [49].

2.2. Modified Mineral Adsorbents

All modifications of mineral adsorbents are aimed at improving their sorption parameters towards particular impurities. These modifications generally include a thermal treatment (calcination, expanding) and the functionalization of the mineral's surface with organic compounds.

Calcinated sorbents (e.g., diatomite) exhibit a more developed specific surface area, a higher sorption capacity (0.50–1.30 g/g) and slightly lower bulk density (0.45–0.60 kg/dm^3). Examples of expanded adsorbents are perlite and vermiculite; these are characterized by a low bulk density (approximately 0.25 kg/dm^3) and good buoyancy, and they can therefore be applied to remove oils from the surface of water [50,51]. Their cost of production is relatively high, and these sorbents tend to return absorbed substances under the influence of external physical and mechanical factors [33].

In recent years, the modification of the surface of minerals with surfactants from the group of quaternary ammonium salts has become a very popular field of research [52–56]. This kind of functionalization changes the character of a mineral materials' surface from hydrophilic to hydrophobic, which usually increases the affinity of the sorbent for organic pollutants. One of the most common organic modifiers is hexadecyltrimethylammonium bromide (HDTMA-Br), typically used to modify the surface of clay minerals. Attempts have been made relatively recently to use these compounds for the modification of natural and synthetic zeolites [57–60]. HDTMA-Br and other salts of this kind can bond to mineral surfaces due to electrostatic forces through exchange with sodium, potassium or calcium cations. These organo-minerals have been studied in terms of the removal of petroleum products in the form of oils, fuels and their derivatives, such as BTEX compounds [52,61–63].

The process of modifying mineral materials with organic compounds increases the cost of the final sorbent, and thus far, this type of organo-mineral sorbents has not found a wide industrial application. It is noteworthy that problems may arise with their further disposal and utilization, as well as the risks related to the toxicity of quaternary ammonium salts [64].

3. Overview of the Sorption of Petroleum Substances by Mineral Adsorbents

3.1. Sorption of Land-Based Oil Spills

The most popular method of sorption capacity (*SC*, g/g) determination are weighing the sorbent before and after saturation with the petroleum substance. Sorption capacity is calculated using the equation:

$$SC = (M - M_0)/M_0 \tag{1}$$

where M is the weight of the sorbent after sorption of petroleum product, and M_0 is weight of the sorbent before sorption. Literature reports different approaches of *SC* determination via weighing methods. Experimental conditions applied by several researches are described below.

Michel [65] investigated sorption capacity of chalcedonite (Teofilów deposit, Poland), clinoptilolite (Nižný Hrabovec, Zeocem, Slovakia), diatomite (Jawornik deposit, Poland) and quartz sand of two different fractions, 0.5–0.8 and 1.25–2.00 mm, with respect to rape oil and diesel oil. Each of the sorbent (50 g) was placed in the glass columns (d = 3 cm), then was dropped by the oil until the saturation of whole bed and left to drain. *SC* was determined by weighing. Bandura et al. [66] used similar procedure to determine *SC* of clinoptilolite (Sokirnica Mine, Ukraine), diatomite (commercially available sorbent Absodan), and zeolites synthetized from fly ash (Na-P1 and Na-X type) towards diesel and biodiesel oils. Ten grams of the tested sorbent was placed in the glass columns (d = 1 cm), dropped by the oil and left to the bed saturation. Oil excess was slightly removed using pump. The fractions of clinoptilolite and diatomite was 0.5–1.0 mm, whereas zeolites from fly ash was in the powder form.

Muir and Bajda [60] investigated zeolites and organo-zeolites (in the form of powder) in terms of petrol, diesel, engine oil, and used engine oil sorption. Clinoptilolite (Sokirnica Mine) and Na-P1 from fly ash were modified with 8 different surfactants: octadecyltrimethylammonium bromide (ODTMA), hexadecyltrimethylammonium bromide (HDTMA), tetradecyltrimethylammonium bromide (TDTMA), dodecyltrimethylammonium bromide (DDTMA), dioctadecyldimethylammonium bromide (DODDMA), dihexadecyldimethylammonium bromide (DHDDMA), ditetradecyldimethylammonium bromide (DTDDMA) and didodecyldimethylammonium bromide (DDDDMA) which belong to quaternary ammonium salts. Each sample was dropped by the oil gradually, until the point of maximum saturation was achieved. After sorption the unmodified adsorbents have been regenerated by burning with sorbed oil. After cooling, the samples were used as adsorbents again, up to 10 times. It was noted that the sorption efficiency after each sorption-combustion-sorption cycle remained similar to the initial material.

Carmody and Frost [52] used different reference sorbents (mineral and natural organic, commonly available and used for petroleum spills cleanup in a form of granules and/or powders), and organo-clays (montmorillonite and bentonite modified with ODTMA and DDDDMA) for sorption of diesel, hydraulic oil and engine oil. The procedure of *SC* determination was based on the ASTM F726-99: Standard Test Method for Sorbent Performance Of Adsorbents [67]. The sample of a sorbent were placed in a basket and dipped into the oil bath for 15 min. Then, the samples were removed from oil bath, left to drain and weighted.

Zhao et al. [68] used exfoliated vermiculite (EV) and its composite with carbon nanotubes (CNT), e.g., (EV/CNT) for the sorption of diesel and soybean oil and applied similar procedure as Carmody and Frost. Known amount of a sorbent was placed in the glass beaker covered with a metal sieve and was poured with the oil. After 5 min the beaker was placed upside down and left to drain. The sorbents exhibited irregular elongated forms of particles/grains (EV 1–2 mm; EV/CNT 15–20 mm).

Ankowski [69] investigated the sorption capacity (*SC*) of commercial sorbents (based on diatomaceous earth and cellulose, provided by Reo Amos, Poland), as well as zeolites from fly ash granulated with the use of bentonite as a binder. The adsorbents were in a granular form and diesel oil was used as a petroleum substance. Measuring cones filled with sorbent were placed in oil bath for 10 min, then collected, drained and weighted.

Alternatively to the weighing, some researchers propose other ways of sorption capacity determination using instrumental methods. These seems to be more convenient in case of powder sorbents, where draining or filtration might be difficult because of the small grain size and the loss of some amount of the material. Sakthviel et al. [56] investigated sorption performance of zeolites and organo-zeolites towards paraffinic process oil. The zeolites was obtained from fly ash, and then functionalized by silanization using propyl-, octyl-, octadecyl-trimethoxysilane and esterification using stearic acid. The sorbents (in a powder form) were mixed with oil in a beaker, stirred for 30 min and then filtrated under vacuum using Buchner funnel for 10 min. The sorption capacity was determined thermogravimetrically (TGA). Zadaka-Amir et al. [70] and Bandura et al. [71] determined *SC* of mineral sorbents towards oils using element analysis, e.g., CHNSO and CHN, respectively, which allows the determination of the carbon content in the samples. Zadaka-Amir et al. used hydraulic oil as a petroleum substance and sepiolite, talc, sand and organo-clays (ODTMA- and

PTMA (phenyltrimethylammonium bromide)-montmorillonite) as the adsorbents (the adsorbents were powders, apart from sand of grain size 0.8–1.5 mm). Bandura et al. investigated *SC* of clinoptilolite (Sokirnica Mine), diatomite and synthetic zeolites from fly ash (Na-P1, Na-X) (all adsorbents in a powder form) towards diesel, biodiesel and used engine oil. In both papers, proposed experimental conditions reflected the practical use of oil spill sorbents. Oil strains were covered by dry sorbents, then collected after oil sorption and analysed via element analysis in terms of carbon content.

After sorption of petroleum substances by minerals, there occur a problem with generated wastes. Besides thermal regeneration [60], spent sorbents can be used as a substrate for the production of lightweight aggregates [72,73]. The utilization of spent adsorbents in this kind of application provides obtained aggregates with new properties, including higher porosity.

Sorption capacities of mineral and organo-mineral adsorbents in the above mentioned research papers are summarized in Table 1.

Table 1. Sorption capacities (*SC*) of mineral and organo-mineral adsorbents towards petroleum substances in the form of oils.

Sorbent	*SC* (g/g)	S_{BET} (m^2/g)	Source
Zeolites			
clinoptilolite (Sokirnica Mine)	0.47–0.65 0.23–0.38	15.88 18.3	[60] [66,71]
clinoptilolite (Nižný Hrabovec)	0.19–0.22	29	[65]
Na-X	0.75–0.79 0.91–1.13	236.4	[71] [66]
Na-P1	0.86–0.91 1.24–1.40 0.89–1.18	75.6 74.9	[71] [66] [60]
zeolites from fly ashes	0.6–0.9	-	[52]
zeolites X	0.37, 1.33	40, 404	[56]
Clay minerals			
vermiculite	1.3	-	[68]
sepiolite	0.97–1.2	258	[70]
talc	0.33	17	[70]
Silica rocks			
diatomite	0.17–0.26	30	[65]
chalcedonite	1.15–1.18	3	[65]
quartz sand	0.2–0.3 0.03–0.05 0.17	- - -	[52] [65] [70]
Modified minerals/organo-minerals			
PTMA-montmorillonite	0.30	-	[70]
ODTMA-montmorillonite	0.37 1.2–1.6	- -	[70] [52]
DDDMA-monmorillonite	3.6–5.2	-	[52]
DDDMA-bentonite	2.1–3.5	-	[52]
zeolite X modified by propyl-, octyl-, octadecyl-trimethoxysilane and stearic acid	1.10, 1.02, 0.86, 1.15	-	[56]
Na-P1 from fly ash and clinoptilolite, both modified with ODTMA, HDTMA, TDTMA, DDTMA, DODDMA, DHDDMA, DTDDMA, DDDDMA	0.80–1.19 0.36–0.75	20–64 4.7–7.0	[60]
Commercial mineral adsorbents			
sodium aluminosilicate	0.27–0.43	-	[69]
Eco-Dry (based on diatomaceous earth)	1.11	-	[69]
Absodan (diatomite)	0.80–0.89 0.41–0.52	24	[66] [71]

Within the group of zeolitic adsorbents, the highest sorption capacities towards oils were noted for synthetic zeolites obtained from fly ash. Depending on the type of structure and surface area, *SC* values were in the range 0.6–1.21 g/g. Lower values of *SC* were characteristic for zeolites granulated with clay and natural zeolite (clinoptilolite). Within the group of clay adsorbents, higher *SC* was reported for vermiculite and sepiolite (0.97–1.3 g/g). Sorption capacity of sepiolite increased from 0.98 to 1.20 g/g after thermal treatment at 300 °C. The authors explain that such improvement might be a result of water loss from the clay which increases the hydrophobicity and affinity for oil. However, thermal treatment at higher temperature (400 °C) slightly decreased *SC* that is connected with some structural changes of the clay. Talc shows a *SC* which is clearly lower than the other adsorbents.

Among the silica adsorbents, diatomites exhibit a great diversity in terms of sorption capacities. This may arise from their different particle sizes or from any thermal treatment provided by manufacturer. In addition, diatomites may have a different mineral composition depending on their geological occurrence, which directly influences their sorption properties. Lower sorption parameters are clearly shown by quartz sand; however, due to its widespread availability and very low cost, it is sometimes used for the removal of large oil spills on rigid/paved surfaces.

Grained minerals can be also used for the removal of oil spills from water surfaces, especially near coastal areas. Boglaienko and Tansel [18,74–77] used sand, limestone and clay for the removal of floating oils using a separation technique. The addition of granular particles into the hydrophobic floating phase gradually increases its density. Due to the cohesive forces, the oil phase covers the mineral particles, creating "particle-oil aggregates". As the density of these aggregates reaches a critical point, they separate from the floating organic phase and sink under gravity. In the case of fine quartz sand, the efficiency of removal of oil (% w/w) increased from 31.44% to 94.35% with an increase of the amount of granular particles added. The experimental conditions were as follows: 1 mL crude oil; 100 mL water; and 0.5, 1.0, 1.5, 2.0, and 3.5 g of sand. This method allows control over the mobility of oil spills on a water surface, thus reducing the negative impact of petroleum substances on aquatic environments.

The modification of minerals using organic compounds influences their sorption abilities towards oils in various ways, depending on the essential properties of the mineral (its type, surface area, cation exchange capacity CEC, external cation exchange capacity ECEC, etc.), on the properties of the organic compound (the length and multiplicity of an organic chain), and lastly on the process and efficiency of the incorporation of the organic compound into the mineral structure. As a rule, the effect is positive in the case of clay minerals. This is due to the fact that the hydrophilic properties of the surface of a clay mineral change to hydrophobic, and the interlayer spaces within the structure of these minerals increase as a result of organic compound incorporation. On the other hand, in the case of the organo-zeolites investigated, sorption capacity towards oils decreased. This kind of surface modification blocks the pores of zeolites, thus reducing their specific surface area and limiting the access of oil into the pores [56,60].

The recent literature reports that the sorption of petroleum products on mineral adsorbents is affected by factors such as the textural properties of the adsorbent (specific surface area, the contribution and surface of mesopores, mesopore diameters), particle size distribution and bulk density of the adsorbent, and the density and viscosity of an adsorbate [71]. Low bulk density and fine particles favour the formation of capillaries between the material grains, and the contact area between the grains increases with a decrease in their diameter. This leads to an increase in the surface area available for petroleum substances. It can be assumed that the sorption of oil products on porous mineral adsorbents occurs via two mechanisms. The first is associated with the filling of the available pores and capillaries between the sorbent grains using capillary action (mass transfer/flow). Capillary action depends on the effective diameter of the capillary, the surface energy of the interior wall of the capillary, and the viscosity of the oil. The second mechanism is sorption on the outer surface of the sorbent through the formation of an oil layer around the sorbent grains (or optionally around the agglomerates of grains). As in the case of surface adsorption on quartz sand [78], oil can create a uniform layer (film) or irregular clusters, depending on the morphology of the surface, its irregularities and roughness, and the properties of the oil. Adsorbents with a well-developed mesoporous structure exhibit relatively higher *SC* in relation to oils. Micropores present in the structure of minerals are unavailable for large oil particles [65,71]. In addition, *SC* is generally higher for oils of higher density and viscosity [36,37,79]. This phenomenon can be explained by the fact that the substances with a higher molecular weight tend to be preferentially adsorbed over those of lower molecular weight [80]. Generally, long-chain hydrocarbons of relatively high molecular weight dominate the chemical composition of higher-density oils (such as spent engine oil).

3.2. Sorption of Benzene and Its Derivatives Present in Aqueous Solutions by Organo-Mineral Adsorbents

The issue of the removal of petroleum compounds such as benzene, toluene, ethylbenzene, xylenes (BTEX), other benzene derivatives and volatile organic compounds from water is widely discussed in the literature. Mineral and organo-mineral sorbents are proposed as the alternative to activated carbon or organic polymers. Most studies concern the sorption of BTEX as volatile petroleum derivatives representatives [80–87]. In these papers, different organic compounds have been used for mineral surface functionalization, and the influences of pH, adsorbent dosage and initial concentration on sorption efficiency have been considered, together with the equilibrium and kinetics of sorption. The authors have explored sorption capacities (q_e, mg/g) of the adsorbents using standard batch technique and applied popular isotherm and kinetic models for the evaluation of possible sorption mechanisms. The concentrations of organic compounds were determined using chromatography methods such as GC-FID (Gas Chromatography coupled to a Flame Ionization Detector) [49,81,87], GC-MS (Gas Chromatograph coupled to a Mass Spectrometer) [57], HPLC (High Performance Liquid Chromatography) [80]. Table 2 summarizes the results obtained for BTEX and other organic compounds sorption on natural and modified minerals including kinetic and isotherm parameters of the best fitting.

The pseudo-first-order model is described by the equation:

$$q_t = q_e(1 - e^{-k_1 t}) \tag{2}$$

where k_1 is the pseudo-first order rate constant of adsorption (L/min), q_e (mg/g) and q_t (mg/g) are the quantities of the adsorbed substance at equilibrium and at time t respectively. This equation is often used to interpret experimental data; however, in practice it shows large deviations in fitting procedures.

The pseudo-second-order model [88] is currently most widely used kinetic equation:

$$q_t = (q_e^2 k_2 t)/(1 - q_e k_2 t) \tag{3}$$

where k_2 is the pseudo-second-order rate constant of adsorption (g/(mg·min)).

The intraparticle diffusion model (Weber and Morris) is expressed by the following equation:

$$q_t = K_{id} t^{0.5} + C \tag{4}$$

and assumes that the adsorption varies linearly with the square of the contact time. K_{id} is the rate constant of intraparticle diffusion (mg/(g·min$^{0.5}$)), $t^{0.5}$ is the square root of the time, and C is the intercept related to the thickness of the boundary layer.

Table 2. Sorption capacities (q_e) of selected adsorbents towards volatile organic compounds present in aqueous solutions and the most suitable kinetic and isotherm models with parameters.

Sorbent	S_{BET} (m²/g)	Adsorbate	T (°C)	q_e (mg/g)	C (mg/L)	Kinetic Model	k Parameter	Isotherm Model	Isotherm Constants		Source
									K_f	1/n	
diatomite raw	38.4	benzene	20	0.10	50	Pseudo-second-order, k_2 (g·(mg·h)$^{-1}$)	3.3263	Freundlich	3.68×10^{-5}	2.47	[49]
		toluene		0.15	50		1.8276		1.92×10^{-3}	1.33	
		ethylbenzene		0.30	50		0.9134		4.44×10^{-2}	0.55	
		p-xylene		0.34	50		0.9214		6.11×10^{-2}	0.48	
		o-xylene		0.20	50		0.5226		4.6×10^{-2}	0.46	
		MTBE		0.08	100		2.1956		5.92×10^{-10}	4.38	
diatomite calcinated at 550 °C	43.3	benzene	20	0.06	50	Pseudo-second-order, k_2 (g·(mg·h)$^{-1}$)	18.0996	Freundlich	1.55×10^{-13}	8.13	[49]
		toluene		0.21	50		7.2447		6.61×10^{-12}	7.29	
		ethylbenzene		0.62	50		4.7319		7.41×10^{-11}	6.07	
		p-xylene		0.82	50		2.1946		2.74×10^{-10}	5.71	
		o-xylene		0.64	50		2.6614		6.02×10^{-13}	7.25	
		MTBE		0.01	100		48.9326		4.36×10^{-11}	4.82	
diatomite calcinated at 750 °C	31.8	benzene	20	0.21	50	Pseudo-second-order, k_2 (g·(mg·h)$^{-1}$)	0.3069	Freundlich	1.35×10^{-7}	4.45	[49]
		toluene		0.22	50		0.1868		9.82×10^{-9}	5.50	
		ethylbenzene		0.58	50		0.0557		1.95×10^{-6}	3.63	
		p-xylene		0.75	50		0.1581		3.18×10^{-6}	3.46	
		o-xylene		0.50	50		0.3335		1.02×10^{-9}	5.46	
		MTBE		0.21	100		0.4441		48.42	1.53	
diatomite calcinated at 950 °C	7.7	benzene	20	0.10	-	Pseudo-second-order, k_2 (g·(mg·h)$^{-1}$)	2.0324		-	-	[49]
		toluene		0.11			0.7896				
		ethylbenzene		0.38			0.2091				
		p-xylene		0.42			0.5710				
		o-xylene		0.23			1.2817				
		MTBE		0.05			0.0002				

Table 2. *Cont.*

Sorbent	S_{BET} (m²/g)	Adsorbate	T (°C)	q_e (mg/g)	C (mg/L)	Kinetic Model	k Parameter	Isotherm Model	Isotherm Constants		Source
									K_f	1/n	
HDTMA smectite	61.7	benzene	23	0.52	10.82	Langmuir-Freundlich, $k_a \times 10^2$ $[L^n/(g{\cdot}min{\cdot}mg^{n-1})]$	5.19	-	-	-	[80]
		toluene		0.69	29.06		9.30				
		ethylbenzene		0.72	8.58		5.04				
		p-xylene		0.75	8.55		6.05				
		m-xylene		0.76	8.52		5.97				
		phenol		0.51	10.00		0.34				
									K_f	1/n	
PEG montmorillonite	27.8	benzene	25	3.47	37.5	Pseudo-second-order, k_2 (g·(mg·h)⁻¹)	0.264	Freundlich	0.024	0.763	[81]
		toluene		4.18	37.5		0.241		0.043	0.725	
		ethylbenzene		5.12	37.5		0.199		0.027	0.637	
		xylene		6.00	37.5		0.153		0.016	0.719	
HDTMA clinoptilolite	-	benzene	20	0.22	9	Pseudo-second-order, k_2 (g·(mg·h)⁻¹)	4.46	-	-		[87]
		toluene		0.20	9		6.49				
		ethylbenzene		0.16	9		5.60				
		xylene		0.13	9		4.25				
CPB clinoptilolite	-	benzene	20	0.35	9	Pseudo-second-order, k_2 (g·(mg·h)⁻¹)	1.94	-	-		[87]
		toluene		0.30	9		1.64				
		ethylbenzene		0.27	9		1.47				
		xylene		0.26	9		1.42				
HDTMA clinoptilolite granulated	-	benzene	20	1.07	9	Pseudo-second-order, k_2 (g·(mg·h)⁻¹)	5.12	-	-		[87]
		toluene		0.83	9		2.87				
		ethylbenzene		0.67	9		6.46				
		xylene		0.62	9		8.82				
CPB clinoptilolite granulated	-	benzene	20	1.16	9	Pseudo-second-order, k_2 (g·(mg·h)⁻¹)	1.94	-	-		[87]
		toluene		0.89	9		4.46				
		ethylbenzene		0.73	9		7.27				
		xylene		0.66	9		4.88				

Table 2. *Cont.*

Sorbent	S_{BET} (m²/g)	Adsorbate	T (°C)	q_e (mg/g)	C (mg/L)	Kinetic Model	k Parameter	Isotherm Model	Isotherm Constants		Source
									K_f	$1/n$	
Zeolite Y	650.3	benzene	20	0.01	10	Pseudo-second-order, k_2 (g·(mg·h)$^{-1}$)	-	-	-		[57]
		toluene		0.10	10						
		ethylbenzene		0.05	10						
		o-xylene		0.10	10						
		m,p-xylene		0.02	10						
									K_t	b	
HDTMA-Y 100	-	benzene	28	12.13	10	Pseudo-second-order, k_2 (g·(mg·h)$^{-1}$)	7.114	Temkin	0.463	2.848	[57]
		toluene		13.75	10		3.825		0.368	1.362	
		ethylbenzene		13.86	10		1.891		5.125	0.429	
		o-xylene		13.98	10		1.460		0.869	0.554	
		m,p-xylene		13.98	10		2.092		0.924	0.694	
Na-P1	88	benzene	20	1.38–6.11	-	-	-	-	-		[58]
		toluene		1.44–8.64							
		p-xylene		1.81–11.41							
HDTMA-NaP1	-	benzene	20	1.36–10.28	-	-	-	-	-		[58]
		toluene		1.40–9.99							
		p-xylene		1.79–11.47							

MTBE: methyl tertiary butyl ether, PEG: poly(ethylene glycol) butyl ether, CPB: *n*-cetyl pridinium bromide, HDTMA-Y 100: zeolite Y modified with HDTMA in loading level 100% of CEC; k_2: pseudo-second order rate constant, T (°C): sorption temperature, C (mg/L): initial concentration of the adsorbate in the kinetic study, k_a: Langmuir-Freundlich rate constant, K_f: Freundlich constant related to the saturation capacity (mg·kg^{-1} (mg/L)^n), $1/n$: Freundlich constant related to intensity of the adsorption, K_t: Temkin constant (L/g), b: Temkin constant (kJ/mol) related to the heat of sorption.

The Langmuir isotherm describes the adsorption at specific homogeneous sites on the adsorbent surface, without significant interaction between the adsorbed species. The Langmuir isotherm is represented by the following equation:

$$q_e = (q_m K_L C_e)/(1 + K_L C_e) \tag{5}$$

where C_e is the solute concentration at equilibrium (mg/L), q_e is the adsorption capacity at equilibrium (mg/g), K_L is the Langmuir adsorption constant (L/mg), and q_m is monolayer capacity (mg/g).

The Freundlich model can be applied to non-ideal systems, including multilayer adsorption processes on heterogeneous surfaces, and is expressed by the following equation:

$$q_e = K_F C_e^{1/n} \tag{6}$$

where K_F (mg·kg^{-1} (mg/L)^n) and $1/n$ are the Freundlich adsorption isotherm constants related to the saturation capacity and intensity of adsorption respectively. Values of $1/n$ of between 0.1 and 1 indicate favorable adsorption. Higher values of $1/n$ suggest unfavorable conditions for adsorption and the possibility of competitive adsorption.

The Langmuir-Freundlich model can be described as:

$$q_e = q_m (K_a C_e)^n / (K_a C_e)^n + 1 \tag{7}$$

where: q_e is the adsorption capacity, q_m is the adsorption capacity of the system (mg/g), C_e is the aqueous phase concentration at equilibrium (mg/L), described as previously and K_a is the affinity constant for adsorption (L^n/g·min·mg^{n-1}), n is the index of heterogeneity.

The Temkin isotherm considers the effects of indirect adsorbate–adsorbate interactions and assumes that heats of adsorption will more often decrease than increase with increasing coverage. The equation has the form:

$$q_e = (RT/b) \ln(AC_e) \tag{8}$$

where T is the absolute temperature (K), R is the gas constant, A is the Temkin isotherm constant (L/g), b is the Temkin constant related to the heat of sorption (kJ/mol).

The detailed description on the experimental conditions connected with the data presented in Table 2 were added below.

Aivalioti et al. [49] studied the sorption of benzene, ethylbenzene, toluene, xylenes and methyl tertiary butyl ether (MTBE) from aqueous solutions on diatomite (Kazani, Northern Greece) [49]. The mineral was milled and the fraction of 64–120 μm was chosen for the study. The diatomite was thermally activated at 550, 750 and 950 °C. The sorption kinetic and isotherm experiments was conducted for multi-component solution (a mixture of BTEX and MTBE) at 20 °C. The maximum duration of contact time was 21 days but the time needed to achieve equilibrium state was 10 days. The initial concentration of the multi-component solutions for kinetic and equilibrium studies was around 250 mg/L of BTEX (50 mg/L of each compound) and 100 mg/L of MTBE. Kinetic study was performed using diatomite to solution ratio of 1:10, whereas the sorbent to solution ratios in case of equilibrium study were 1:25, 1:15, 1:10, 1:5, 1:4, and 1:3. The sorption of BTEX compounds at equilibrium state was in the range 0.1–0.38 mg/g for raw diatomite (without thermal activation) and increased to 0.8 mg/g for diatomite calcinated at 750 °C. The uptake of MTBE was relatively low (0.01–0.21 mg/g). Calcination provided loss of water and impurities from pores and active sites of the mineral, thus increasing the external surface area available for adsorbate molecules. Based on the correlation coefficients R^2 values the experimental data fit the pseudo-second order kinetic model and Freundlich isotherm model. The values of Freundlich constant $1/n$ higher than 1 suggest unfavorable sorption. Sorption capacities followed the order: *p*-xylene > ethyl-benzene > *o*-xylene > toluene > benzene > MTBE resulted from the descending order of hydrophobicity (based on their octanol–water coefficient log values: 3.15, 3.15, 2.77, 2.69, 2.13 and 1.06, respectively), molecular weight, and approximately ascending order of water solubility (198, 152, 175, 515, 1700 and 47,000 mg/L, respectively) [49].

Carvalho et al. [80] used smectite (Paraiba, Brazil) functionalized with HDTMA for the sorption of benzene, toluene, ethylbenzene, *p*-xylene, *m*-xylene and phenol from single-component and multi-component solutions at 23 °C. The influence of pH (4, 7 and 9), contact time (time intervals in the range 0.5–240 min) and initial concentration of the solution (from 2 to 29 mg/L at optimal pH = 9 and contact time 240 min) on sorption efficiency was investigated. The most efficient sorption levels were noted at pH = 9. At lower pH values, the ions H^+ might coordinate to active sites (surface silanol groups) of the organoclay and block them. For single-component system the removal efficiencies ranged between 55% and 90%, whereas those for multi-component system were lower (between 30% and 90%). Sorption capacity (q_e) towards BTEX was in the range 0.25–0.4 mg/g for multi-component system (Table 2) and followed the order xylenes > ethylbenzene > toluene > benzene. The adsorption equilibrium was established after 60 min. Authors applied pseudo-first, pseudo-second and Langmuir-Freundlich models for kinetic data interpretation and the best fitting was noted for Langmuir-Freundlich. Analyzing the obtained kinetic parameters they have assumed that: (i) toluene and *p*-xylene exhibit a high adsorption affinity to organoclay surface and high binding energy of adsorption; (ii) phenol and ethylbenzene cover many active sites of organoclay and that they are not selective for one particular active site; (iii) toluene have the highest rate of adsorption, whereas phenol have the slowest rate.

Nourmoradi et al. [81] examined montmorillonite (Laviosa Co., Livorno, Italy) modified with poly ethylene glycol (PEG) in terms of benzene, toluene, ethylbenzene and xylenes sorption from aqueous solution (multi-component). Grain size of the sorbent did not exceed 125 μm. Loading rates of the surfactant (0.2–4.00 CEC of the montmorillonite), contact time (0–24 h), pH, adsorbate initial concentration C_0 (10–200 mg/L of BTEX mixture) and solution ion strength (20–100 mg/L of calcium ions) on the BTEX sorption capacity was investigated at 25 °C. In the studied BTEX solutions C_0 of benzene, toluene and xylene (in mg/L) were equal. Also the influence of temperature (10–40 °C) on sorption efficiency was studied. The regeneration of the adsorbent was performed by heating at 150 °C at 5, 10 and 20 min, and the efficiency of sorption for each regeneration cycle was further estimated. The optimal surfactant loading rate was 2.0 CEC which provided the highest BTEX uptake. Higher loading rates resulted in the decrease of BTEX sorption. pH and ion strength did not influenced significantly the BTEX adsorption. Sorption increased rapidly during the first hour of sorption and the equilibrium was achieved at the contact time of 24 h. The adsorption capacity (q_e) was in the range 3.47–5.12 mg/g and followed the order: xylene > ethylbenzene > toluene > benzene. The experimental data were analysed using pseudo-first-order, pseudo-second-order, and intraparticle diffusion kinetic models and Langmuir, Freundlich and D–R isotherm models. The best fitting was obtained for pseudo-second-order model and Freundlich isotherm. Obtained parameters indicate the highest sorption rate for xylenes, and the favorable physical sorption ($1/n < 1$). The thermodynamic study revealed that the BTEX sorption by PEG-montmorillonite was spontaneous, endothermic and favorable at higher temperatures. Sorption efficiency of regenerated PEG-montmorillonite was 18.14%–23.08% and 51.28%–60.70% of the original material after 5 and 20 min of heating, respectively.

Seifi et al. [87] used clinoptilolite (Miyaneh, Iran) modified with HDTMA and CPB (*n*-cetyl piridinium bromide) for the sorption of BTEX compounds from multi-component solution. The clinoptilolite was in powder (590–840 μm) and granulated forms before modification (the granulation process is described in the paper). The paper presents a comprehensive study of the sorption kinetics. Batch experiments for the single-component systems were carried out at 20 °C within the contact time range 0.5–72 h. The initial concentration of each compound was 9 mg/L. The optimal time for achieving the equilibrium was 24 h for natural and 8 h for granulated adsorbents. Intra-particle diffusion models (Weber and Morris and Vermeulen) and surface reaction models (pseudo-first order, pseudo-second order, and Elovich) were applied for interpretation the experimental data and pseudo-second order model was the most suitable. It was noticed that the intra-particle diffusion is prevailing in the first stage of adsorption for a relatively short time. Sorption capacities for organozeolites were in the range 0.15–0.35 and 0.6–1.2 mg/g for granulated nano-organozeolites. The sorption capacities followed the order: benzene > toluene > ethylbenzene > xylenes. The adsorbents modified by CPB showed higher adsorption capacity compared to the HDTMA-modified adsorbents.

Vidal et al. [57] used commercial synthetic zeolite Y and HDMA as a surfactant for zeolite modification. The HDTMA loading level was 0.5, 1.0 and 2.0 of the total CEC of the zeolite Y. The best results were obtained at an HDTMA loading level of 1.0 of the zeolite CEC. Kinetics and isotherm studies for multi-component BTEX solutions were performed at 28 °C. The contact time ranged from 30 min to 48 h and the BTEX concentration was 10 mg/L for each compound. The sorption equilibrium was reached within 6 h. In the isotherm study, BTEX initial concentration for each compound was in the range 2–60 mg/L and 12–360 mg/L for multi-component solution. The experimental data was evaluated using pseudo-first, pseudo-second and intraparticle diffusion kinetic models and Langmuir, Freundlich, Redlich-Peterson and Temkin isotherms. Based on the values of correlation coefficients the best fitting was obtained for pseudo-second order model and Temkin isotherm indicating the effects of indirect adsorbate-adsorbate interactions. Adsorption capacity was 12.13 mg/g for benzene, 13.75 mg/g for toluene, 13.86 mg/g for ethylbenzene, 13.98 mg/g for *o*-xylene, and 13.98 mg/g for *m*- and *p*-xylene. Sorption capacity for unmodified zeolite was significantly lower. The adsorption capacities followed the order: *m*-, *p*-xylene > *o*-xylene > ethylbenzene > toluene > benzene, likewise in other papers. Regeneration of the HDTMA-zeolite Y were performed by heating the sorbent at 100 °C for 6 h, followed by a new adsorption cycle (up to 5 times). The results have shown that the sorption efficiency was at the same level after 4 cycles of regeneration, except benzene, for which sorption efficiency has been reduced after first regeneration cycle.

Szala et al. [58] used zeolite Na-P1 prepared from fly ash modified with HDTMA using loading levels of 0.2, 0.4, 0.6, 0.8 and 1.0 of the external cation exchange capacity (ECEC) of Na-P1 for the sorption of benzene, toluene and xylene. Sorption experiments were carried out at room temperature (20 °C) at the initial concentration. Depending on the initial concentration of benzene, toluene and xylene (50, 100, 250 mg/L), the sorption capacities varied from about 1.38 to 11.47 mg/g. The best adsorption uptake was achieved for HDTMA-Na-P1 with loading level of 1.0 ECEC. The modification improved the sorption capacity of Na-P1 in terms of BTX compounds (benzene, toluene, xylenes) by several per cent.

The sorption selectivity in most studies followed the order xylenes > ethylbenzene > toluene > benzene [57,80–82]. The dependency might be explained by molecular weight of the adsorbates, their hydrophobicity, water solubility, the size and geometrical shape of adsorbate particles, and the type of functional groups. Generally, the adsorbates of higher molecular weight and larger sizes are adsorbed more preferably. The opposite tendency was observed by Seifi et al. [87]. The studies revealed that the functionalization of the surface of mineral adsorbents by surfactants can improve the efficiency of BTEX sorption from aqueous solution. Different loading levels of surfactant provide various sorption efficiency. An optimal loading level was 1.0–2.0 of CEC and higher loading level limited sorption efficiency.

The lowest values of sorption capacities were obtained for diatomite, zeolite Y and smectite modified by HDTMA. The most efficient adsorbents were montmorillonite modified with PEG and zeolite Y modified with HDTMA. Most presented research revealed that sorption kinetics of organic compounds adsorption followed the mechanism of pseudo-second order model. The rate constant of pseudo-second kinetic, k_2, reflects the adsorption rate which was the lowest for diatomite. Indeed, in case of this mineral, the equilibrium state of adsorption was achieved after 10 days. Studied organo-mineral adsorbents exhibited faster rate of the organic compounds adsorption, confirmed by the experimental results and the calculated rate constants. In general, the adsorption was rapid in the first stage. The following stage was a slower adsorption process where the increase of the adsorption capacity became much slower than that of first stage, followed by the state of equilibrium in which the adsorption capacity remained at a constant level. The most efficient adsorbent in terms of the rate of the adsorption was HDTMA-smectite. The first rapid stage lasted about 10 min and the equilibrium state was achieved after 60 min. Isotherm models used in several studies indicates the high applicability of Freundlich model that assumes the adsorption of physical nature onto heterogeneous surface.

The important issue is regeneration of the used adsorbents since they constitute a waste that can be hazardous for the environment. Results presented by Nourmoradi et al. [81] and Vidal et al. [57] show that it is possible to regenerate the used organo-mineral adsorbents by thermal treatment without significant loss of further sorption efficiency.

3.3. Sorption of Volatile Petroleum Derivatives

For the removal of volatile petroleum derivatives (including BTX), activated carbons, polymers or silica gels of very high surface area are generally used [83,89]. In recent years, natural and synthetic zeolites have also been considered as alternatives to these. Zeolites exhibit great thermal stability in contrast to polymers or carbons. They can therefore be applied instead, in particular conditions such as high temperatures or the risk of fire. Sorption capacities of different zeolites towards volatile organic compounds from gas streams are presented in Table 3. The investigations were carried out using a fixed-bed flow contractor [88–91] and/or TPD techniques (temperature programmed desorption) [92,93].

Table 3. Sorption capacities (q_e) of selected adsorbents towards volatile organic compounds from gas streams.

Sorbent	S_{BET} (m²/g)		Adsorbate	T (°C)	q_e (mg/g)		Source
	623 K	823 K			623 K	823 K	
natural zeolite—NZ	205	170	toluene	20	0.028	0.009	[90]
acid treated NZ	434	369	toluene	20	0.131	0.104	[90]
ion exchanged NZ	181	222	toluene	20	0.035	0.056	[90]
double ion exchanged NZ	171	261	toluene	20	0.082	0.074	[90]
natural mordenite	20		benzene toluene p-xylene	20	3.6 1.9 1.9		[91]
acid-treated mordenite	128		benzene toluene p-xylene	20	21.9 25.8 11.4		[91]
MFI	377		acetone p-xylene n-hexane	25	105.71 135.88 110.31		[92]
*BEA	493		acetone p-xylene n-hexane	25	124.87 124.21 106.00		[92]
STT	536		acetone p-xylene n-hexane	25	141.12 102.98 97.38		[92]
CHA	803		acetone p-xylene n-hexane	25	5.808 4.25 15.51		[92]
faujasite Y HY901	591		toluene MEK	25	1068.82 757.155		[93]
faujasite X MS13X	582		toluene MEK	25	340.92 778.79		[93]

Table 3. *Cont.*

Sorbent	S_{BET} (m²/g)		Adsorbate	T (°C)	q_e (mg/g)		Source
	623 K	823 K			623 K	823 K	
ZSM-5	390		toluene	40	65.5		[94]
			isopentane		32.3		
			ethanol		81.3		
beta zeolite	446		toluene	40	39.5		[94]
			isopentane		4.0		
			ethanol		10.5		
TS-1	414		toluene	40	74.0		[94]
			isopentane		29.5		
			ethanol		15.8		
silicate-1	356		toluene	40	28.9		[94]
			isopentane		11.5		
			ethanol		18.4		
Na-P1 zeolite from fly ash	94.49		benzene	40	0.11		[95]
			toluene		0.97		
			o-xylene		2.78		
			m-xylene		2.76		
			p-xylene		2.18		
Na-X zeolite from fly ash	157.43		benzene	40	29.97		[95]
			toluene		48.38		
			o-xylene		57.88		
			m-xylene		59.86		
			p-xylene		61.88		
clinoptilolite	18.33		benzene	40	0.05		[95]
			toluene		0.27		
			o-xylene		0.77		
			m-xylene		0.41		
			p-xylene		0.30		
diatomite	23.51		benzene	40	0.02		[95]
			toluene		0.09		
			o-xylene		0.10		
			m-xylene		0.10		
			p-xylene		0.10		

MEK—methylethylketone; T (°C)—temperature of adsorption.

Alejandro et al. [90] investigated the sorption of toluene on natural zeolite from Chile (53% clinoptilolite, 40% mordenite and 7% quartz) which was treated with use of hydrochloric acid and ammonium sulphate and then thermally activated at 623 and 823 K (activation was performed for chemically treated and untreated samples). The grain size was 0.3–0.425 mm. The procedure of toluene adsorption-ozonation experiments was carried out in a fixed-bed flow contactor (ID 45 mm) operating at room temperature (20 °C) and 101 kPa. The concentration of toluene was regulated by bubbling dry air into pure liquid toluene and diluted to a desired concentration by mixing with an air stream (22.2 μmol/L). The stream with toluene vapor of constant concentration was continuously supplied over the zeolite bed until saturation was achieved. Next, toluene was replaced by ozone. Toluene concentrations at the inlet and outlet streams were monitored on-line by gas chromatography (GC-FID). Sorption capacities in terms of toluene were 1.46 and 0.55 μmol/m² for zeolite heated at 623 and 823 K respectively. The specific surface area (S_{BET}) of thermally treated zeolite was 205 and 170 m²/g, and calculated sorption expressed in mg/g was 0.028 and 0.009 respectively. Sorption efficiency increased after acid treatment to 3.27 and 3.07 μmol/m² (0.131 and 0.104 mg/g), and after ion exchange to 2.07–5.22 and 2.76–3.09 μmol/m² (0.035–0.082 and 0.056–0.074 mg/g) respectively (higher values were obtained after a double ion exchange process). After these thermal and chemical modifications, both the specific surface area (S_{BET}) of zeolite and its surface character have changed. Higher temperatures caused a decrease in S_{BET} of untreated and acid treated samples, and an increase in the case of zeolite after ion exchange. The surface character was responsible for toluene sorption capacity, and for the efficiency of the oxidative regeneration processes.

Another zeolite which has been used for volatile organic compounds BTX (benzene, toluene and *p*-xylene) sorption is mordenite (natural and after acid treatment) investigated by Valdés et al. [91]. The fraction chosen for the experiments was 0.3–0.425 mm. The sorption experiment was performed as described by Alejandro et al. [90] using fixed-bed flow contactor (4mm ID) at 20 °C and 101 kPa. Pre-adsorbed benzene, toluene or *p*-xylene was then desorbed using a temperature programmed desorption (TPD) procedure by heating the sample up to 550 °C (heating rate of 10 °C/min). The influence of moisture was also studied by mixing a humidified stream (40% of relative humidity RH) with the BTX stream at the inlet. It was found that acid treatment increased S_{BET} significantly, from 20 to 128 m²/g. Simultaneously, sorption efficiency towards benzene, toluene and *p*-xylene increased from 3.6, 1.9 and 1.9 mg/g to 21.9, 25.8 and 11.4 mg/g respectively. In the presence of moisture, the values of adsorption capacities of raw mordenite and after acid treatment towards benzene, toluene and *p*-xylene decreased by 21%, 29%, 32%; and by 68%, 65%, 41%, respectively.

Cosseron et al. [92] used several types of zeosils (pure silica zeolites; chabazite (CHA-structure type), SSZ-23 (STT-structure type) with cage-like structure, silicalite-1 (MFI-structure type) and beta (*BEA-structure type) with channel structure) for the adsorption of *n*-hexane, acetone and *p*-xylene. The measurements were performed under flow using a thermogravimetric balance Setaram TG92 instrument (at relative pressure $p/p_0 = 0.5$ at 25 °C). Sorption capacities of *p*-xylene were in the range 0.04–1.28 mmol/g (4.25–135.88 mg/g) depending on the zeolite structure.

Kim and Ahn [93] investigated several types of commercial synthetic zeolites from a group of mordenite, faujasite X and faujasite X in terms of the sorption of volatile organic compounds (benzene, toluene, *o*-xylene, *m*-xylene, *p*-xylene, methanol, ethanol, isopropanol, and methylethylketone (MEK)) using a continuous flow system under atmospheric pressure. The temperature of adsorption was 25 °C. Vapor concentration was monitored with a gas chromatograph equipped with thermal conductivity detector (GC-TCD). Desorption experiments were performed after sorption by microwave heating (5 °C/min) to 300 or 500 °C. The highest sorption capacities were noted for the faujasite types of zeolites, abbreviated as HY901 and MS13X. In the case of toluene, *SC*s were 1.07 and 0.34 g/g respectively, whereas for mordenites *SC*s were below 0.10 g/g. The BET surface areas of HY901 and MS13X were similar to that of two other faujasites Y however, their sorption capacities towards tested volatile organic compounds were much higher. It indicates that the sorption capacity of volatile organic compounds by zeolites does not depend on the S_{BET}.

Serrano et al. [94] used the TPD method to determine the sorption capacities towards toluene, isopentane, and ethanol of synthetic zeolites synthesized from chemical reagents; sorption capacities were determined to be in the range 25–75 mg/g. The zeolites synthesized by Serrano exhibited high purity and specific surface area.

In our paper [95] we examined diatomite, Ukrainian clinoptilolite and two types of synthetic zeolites from fly ash (Na-P1 and Na-X) in terms of the sorption of benzene, toluene and xylenes from a gas stream using the TPD technique. Adsorption of hydrocarbons was carried out at 45 °C. The effects of hydrocarbon dosing were analyzed at the reactor outlet using the mass spectrometer equipped with an electron multiplier detector. After the complete saturation of the samples the TPD was performed with a heating rate of 10 °C/min and a constant helium flow of 45 cm³/min in the temperature range 45–800 °C. Sorption capacities in terms of BTX for these materials were within the ranges 0.02–0.10, 0.05–0.77, 0.11–2.78 and 29.97–61.88 mg/g respectively. Na-X showed significantly higher sorption than Na-P1 and other minerals used in the study.

In the papers presented above, the authors have generally considered the influence of the textural and chemical properties of mineral adsorbents on the sorption of volatile organic compounds (including petroleum industry derivatives). Most studies concern adsorbents of the zeolite type. It can be postulated that in the sorption of a compound such as BTX, the most important feature is the pore structure of zeolites; better sorption properties were observed for zeolites with ink-bottle pores [92] and zeolites of faujasite structure (HY901 and MS13X, Na-X) [93,95]. Their extremely high sorption performance with respect to the investigated molecules is a result of their relatively high surface area, with a large contribution of the micropores and their specific structure. These kinds of zeolites behave like molecular sieves towards volatile organic compounds and molecules of similar size due to "windows" of 7.4 Å diameter in the micropore structure of faujasite. It can be noticed that sorption capacities of mordenites were higher for benzene, probably because the molecules of benzene are smaller than xylenes and can penetrate into the channels of mordenite easier. Strong relationship between the crystal structure and adsorption capacity was proved by Kim and Ahn [93] and Bandura et al. [95].

4. Conclusions

This paper presents an overview of recent research papers concerning mineral materials, synthetic zeolites and organo-minerals used as adsorbents for petroleum pollutants present in waters, air and spilled on land, occurring as oils, petroleum industry derivatives and volatile compounds. For the practical application, these adsorbents should meet the following criteria: availability, ease of acquisition, costs, good textural parameters, appropriate grain size.

The literature data indicate that minerals with high mesoporosity are recommended for the removal of land-based petroleum spills; the most promising results have been obtained for diatomites, sepiolite and zeolites from fly ash. Additionally, the synthesis of zeolites from fly ash allows to utilize wastes generated in heat and power plants, thus reducing their negative impact on the environment.

The mechanism of oil substances adsorption on porous surface of minerals includes capillary action connected with filling the available pores and oily layer (film) formation on the external surface and around the adsorbent grains. Oil substances cannot penetrate into narrow micropores of mineral adsorbents. The dependency between oil viscosity and density has been also observed. Generally, more viscous and dense oils were adsorbed in higher amounts than light oils by the same adsorbent material.

In the case of applications for water media, one of the main selection criteria is the hydrophobicity of the sorbent surface, which can be achieved from the modification of a mineral with surfactants such as quaternary ammonium salts. These modifications seem to be promising in the case of clay minerals and zeolites. Organo-clays and organo-zeolites exhibit higher sorption performance towards organic compounds in water media than the raw materials. Most studies revealed that the mechanism of organic compounds adsorption from aqueous solution by minerals and organo-minerlas can be described well by Freundlich isotherm which indicates physical sorption on heterogeneous surface. In terms of kinetics, it followed pseudo-second order model in most presented research.

Investigation on the organic pollutants adsorption from gases, the best results were obtained for synthetic zeolites from the group of faujasite. An important factor in those cases was high surface area and the contribution and structure of micropores. Faujasites can act as molecular sieves, selective for volatile organic compounds.

Mineral sorbents have a number of features relevant to their usage in the removal of petroleum compounds. The most favorable are: availability, environmental friendliness, low cost of acquisition and the possibility of recycling. Therefore, it is fully justified to obtain new types of inorganic and/or organic-inorganic structures dedicated to petroleum substances adsorption, with well-defined textural and chemical properties. It is also necessary to identify new/potential directions of spent adsorbents utilization since they constitute waste hazardous for the environment. Apart from being recovered by calcination, they can also be used as additives in the production of building materials such as lightweight aggregates.

Acknowledgments: We acknowledge the financial support from NCBiR within the Project GEKON 2/O2/266818/1/2015.

Author Contributions: Lidia Bandura was responsible for the article conception, collected literature, wrote the manuscript. Wojciech Franus participated in conception, literature collection, writing article parts, substantive comments and care. Agnieszka Woszuk was responsible for data collection on volatile compounds sorption. Dorota Kołodyńska was responsible for data collection on BTEX sorption from water.

Conflicts of Interest: The authors declare no conflict of interest.

References

1. Paulauskienė, T.; Jucikė, I.; Juščenko, N.; Baziukė, D. The use of natural sorbents for spilled crude oil and diesel cleanup from the water surface. *Water Air Soil Pollut.* **2014**, *225*, 1959–1971. [CrossRef]

2. Zhang, J.; Fan, S.; Yang, J.; Du, X.; Li, F.; Hou, H. Petroleum contamination of soil and water, and their effects on vegetables by statistically analyzing entire data set. *Sci. Total Environ.* **2014**, *476–477*, 258–265. [CrossRef] [PubMed]

3. Moro, A.M.; Brucker, N.; Charão, M.F.; Sauer, E.; Freitas, F.; Durgante, J.; Bubols, G.; Campanharo, S.; Linden, R.; Souza, A.P.; et al. Early hematological and immunological alterations in gasoline station attendants exposed to benzene. *Environ. Res.* **2015**, *137*, 349–356. [CrossRef] [PubMed]

4. Aguilera, F.; Méndez, J.; Pásaro, E.; Laffon, B. Review on the effects of exposure to spilled oils on human health. *J. Appl. Toxicol.* **2010**, *30*, 291–301. [CrossRef] [PubMed]

5. Alonso-Alvarez, C.; Pérez, C.; Velando, A. Effects of acute exposure to heavy fuel oil from the Prestige spill on a seabird. *Aquat. Toxicol.* **2007**, *84*, 103–110. [CrossRef] [PubMed]

6. Camacho, M.; Boada, L.D.; Orós, J.; Calabuig, P.; Zumbado, M.; Luzardo, O.P. Comparative study of polycyclic aromatic hydrocarbons (PAHs) in plasma of Eastern Atlantic juvenile and adult nesting loggerhead sea turtles (Caretta caretta). *Mar. Pollut. Bull.* **2012**, *64*, 1974–1980. [CrossRef] [PubMed]

7. Zytner, R.G. Sorption of benzene, toluene, ethylbenzene and xylenes to various media. *J. Hazard. Mater.* **1994**, *38*, 113–126. [CrossRef]

8. Półka, M.; Kukfisz, B.; Wysocki, P.; Polakovic, P.; Kvarcak, M. Efficiency analysis of the sorbents used to adsorb the vapors of petroleum products during rescue and firefighting actions. *Przem. Chem.* **2015**, *1*, 109–113.

9. Chagas-Spinelli, A.C.O.; Kato, M.T.; de Lima, E.S.; Gavazza, S. Bioremediation of a tropical clay soil contaminated with diesel oil. *J. Environ. Manag.* **2012**, *113*, 510–516. [CrossRef] [PubMed]

10. Broje, V.; Keller, A.A. Improved mechanical oil spill recovery using an optimized geometry for the skimmer surface. *Environ. Sci. Technol.* **2006**, *40*, 7914–7918. [CrossRef] [PubMed]

11. Saikia, R.R.; Deka, S. Removal of hydrocarbon from refinery tank bottom sludge employing microbial culture. *Environ. Sci. Pollut. Res. Int.* **2013**, *20*, 9026–9033. [CrossRef] [PubMed]

12. Ji, F.; Li, C.; Dong, X.; Li, Y.; Wang, D. Separation of oil from oily wastewater by sorption and coalescence technique using ethanol grafted polyacrylonitrile. *J. Hazard. Mater.* **2009**, *164*, 1346–1351. [CrossRef] [PubMed]

13. Nikolajsen, K.; Kiwi-Minsker, L.; Renken, A. Structured fixed-bed adsorber based on zeolite/sintered metal fibre for low concentration VOC removal. *Chem. Eng. Res. Des.* **2006**, *84*, 562–568. [CrossRef]

14. Mathur, A.K.; Majumder, C.B.; Chatterjee, S. Combined removal of BTEX in air stream by using mixture of sugar cane bagasse, compost and GAC as biofilter media. *J. Hazard. Mater.* **2007**, *148*, 64–74. [CrossRef] [PubMed]

15. Khan, F.I.; Ghoshal, A.K. Removal of volatile organic compounds from polluted air. *J. Loss Prev. Process Ind.* **2000**, *13*, 527–545. [CrossRef]

16. Al-Majed, A.A.; Adebayo, A.R.; Hossain, M.E. A sustainable approach to controlling oil spills. *J. Environ. Manag.* **2012**, *113*, 213–227. [CrossRef] [PubMed]

17. Fritt-Rasmussen, J.; Brandvik, P.J. Measuring ignitability for in situ burning of oil spills weathered under Arctic conditions: From laboratory studies to large-scale field experiments. *Mar. Pollut. Bull.* **2011**, *62*, 1780–1785. [CrossRef] [PubMed]

18. Boglaienko, D.; Tansel, B. Partitioning of fresh crude oil between floating, dispersed and sediment phases: Effect of exposure order to dispersant and granular materials. *J. Environ. Manag.* **2016**, *175*, 40–45. [CrossRef] [PubMed]

19. Zhao, X.; Liu, W.; Fu, J.; Cai, Z.; O'Reilly, S.E.; Zhao, D. Dispersion, sorption and photodegradation of petroleum hydrocarbons in dispersant-seawater-sediment systems. *Mar. Pollut. Bull.* **2016**, *109*, 526–538. [CrossRef] [PubMed]

20. Cheng, M.; Zeng, G.; Huang, D.; Yang, C.; Lai, C.; Zhang, C.; Liu, Y. Advantages and challenges of Tween 80 surfactant-enhanced technologies for the remediation of soils contaminated with hydrophobic organic compounds. *Chem. Eng. J.* **2017**, *314*, 98–113. [CrossRef]

21. Sundaravadivelu, D.; Suidan, M.T.; Venosa, A.D.; Rosales, P.I. Characterization of solidifiers used for oil spill remediation. *Chemosphere* **2016**, *144*, 1490–1497. [CrossRef] [PubMed]

22. Sundaravadivelu, D.; Suidan, M.T.; Venosa, A.D. Parametric study to determine the effect of temperature on oil solidifier performance and the development of a new empirical correlation for predicting effectiveness. *Mar. Pollut. Bull.* **2015**, *95*, 297–304. [CrossRef] [PubMed]

23. Rosales, P.I.; Suidan, M.T.; Venosa, A.D. A laboratory screening study on the use of solidifiers as a response tool to remove crude oil slicks on seawater. *Chemosphere* **2010**, *80*, 389–395. [CrossRef] [PubMed]

24. Fingas, M. *Review of Solidifiers: An Update 2013*; Prince Williams Sound Regional Citizens' Advisory Council: Edmonton, AB, Canada, 2013.

25. Chebbi, R. Profile of oil spill confined with floating boom. *Chem. Eng. Sci.* **2009**, *64*, 467–473. [CrossRef]

26. Agnello, A.C.; Bagard, M.; van Hullebusch, E.D.; Esposito, G.; Huguenot, D. Comparative bioremediation of heavy metals and petroleum hydrocarbons co-contaminated soil by natural attenuation, phytoremediation, bioaugmentation and bioaugmentation-assisted phytoremediation. *Sci. Total Environ.* **2016**, *563–564*, 693–703. [CrossRef] [PubMed]

27. Chen, M.; Xu, P.; Zeng, G.; Yang, C.; Huang, D.; Zhang, J. Bioremediation of soils contaminated with polycyclic aromatic hydrocarbons, petroleum, pesticides, chlorophenols and heavy metals by composting: Applications, microbes and future research needs. *Biotechnol. Adv.* **2015**, *33*, 745–755. [CrossRef] [PubMed]

28. Wu, M.; Dick, W.A.; Li, W.; Wang, X.; Yang, Q.; Wang, T.; Xu, L.; Zhang, M.; Chen, L. Bioaugmentation and biostimulation of hydrocarbon degradation and the microbial community in a petroleum-contaminated soil. *Int. Biodeterior. Biodegrad.* **2016**, *107*, 158–164. [CrossRef]

29. Atlas, R.M.; Hazen, T.C. Oil biodegradation and bioremediation: A tale of the two worst spills in U.S. history. *Environ. Sci. Technol.* **2011**, *45*, 6709–6715. [CrossRef] [PubMed]

30. Srinivasan, A.; Viraraghavan, T. Oil removal from water using biomaterials. *Bioresour. Technol.* **2010**, *101*, 6594–6600. [CrossRef] [PubMed]

31. Szarlip, P.; Stelmach, W.; Jaromin-Gleń, K.; Bieganowski, A.; Brzezińska, M.; Trembaczowski, A.; Hałas, S.; Łagód, G. Comparison of the dynamics of natural biodegradation of petrol and diesel oil in soil. *Desalin. Water Treat.* **2014**, *52*, 3690–3697. [CrossRef]

32. Adebajo, M.O.; Frost, R.L.; Kloprogge, J.T.; Carmody, O.; Kokot, S. Porous materials for oil spill cleanup: A review of synthesis and absorbing properties. *J. Porous Mater.* **2003**, *10*, 159–170. [CrossRef]

33. Tic, W. Characteristics of adsorbents used to remove petroleum contaminants from soil and wastewater. *Przem. Chem.* **2015**, *1*, 79–84.

34. Wahi, R.; Chuah, L.A.; Choong, T.S.Y.; Ngaini, Z.; Nourouzi, M.M. Oil removal from aqueous state by natural fibrous sorbent: An overview. *Sep. Purif. Technol.* **2013**, *113*, 51–63. [CrossRef]

35. Uzunov, I.; Uzunova, S.; Angelova, D.; Gigova, A. Effects of the pyrolysis process on the oil sorption capacity of rice husk. *J. Anal. Appl. Pyrolysis* **2012**, *98*, 166–176. [CrossRef]

36. Kenes, K.; Yerdos, O.; Zulkhair, M.; Yerlan, D. Study on the effectiveness of thermally treated rice husks for petroleum adsorption. *J. Non-Cryst. Solids* **2012**, *358*, 2964–2969. [CrossRef]

37. Abdullah, M.A.; Rahmah, A.U.; Man, Z. Physicochemical and sorption characteristics of Malaysian *Ceiba pentandra* (L.) Gaertn. as a natural oil sorbent. *J. Hazard. Mater.* **2010**, *177*, 683–691. [CrossRef] [PubMed]

38. Li, J.; Luo, M.; Zhao, C.J.; Li, C.Y.; Wang, W.; Zu, Y.G.; Fu, Y.J. Oil removal from water with yellow horn shell residues treated by ionic liquid. *Bioresour. Technol.* **2013**, *128*, 673–678. [CrossRef] [PubMed]

39. Lin, J.; Shang, Y.; Ding, B.; Yang, J.; Yu, J.; Al-Deyab, S.S. Nanoporous polystyrene fibers for oil spill cleanup. *Mar. Pollut. Bull.* **2012**, *64*, 347–352. [CrossRef] [PubMed]

40. Li, H.; Liu, L.; Yang, F. Hydrophobic modification of polyurethane foam for oil spill cleanup. *Mar. Pollut. Bull.* **2012**, *64*, 1648–1653. [CrossRef] [PubMed]

41. Wu, D.; Fang, L.; Qin, Y.; Wu, W.; Mao, C.; Zhu, H. Oil sorbents with high sorption capacity, oil/water selectivity and reusability for oil spill cleanup. *Mar. Pollut. Bull.* **2014**, *84*, 263–267. [CrossRef] [PubMed]

42. Pichór, W.; Mozgawa, W.; Król, M.; Adamczyk, A. Synthesis of the zeolites on the lightweight aluminosilicate fillers. *Mater. Res. Bull.* **2014**, *49*, 210–215. [CrossRef]

43. Franus, M.; Wdowin, M.; Bandura, L.; Franus, W. Removal of environmental pollutions using zeolites from fly ash: A review. *Fresenius Environ. Bull.* **2015**, *24*, 854–866.

44. Wdowin, M.; Franus, M.; Panek, R.; Badura, L.; Franus, W. The conversion technology of fly ash into zeolites. *Clean Technol. Environ. Policy* **2014**, *16*, 1217–1223. [CrossRef]

45. Wdowin, M. Raw kaolin as a potential material for the synthesis of a-Type Zeolite. *Gospodarka Surowcami Mineralnymi Miner. Resour. Manag.* **2015**, *31*, 45–57.

46. Belviso, C. EMT-type zeolite synthesized from obsidian. *Microporous Mesoporous Mater.* **2016**, *226*, 325–330. [CrossRef]

47. Burham, N.; Sayed, M. Adsorption behavior of Cd^{2+} and Zn^{2+} onto natural Egyptian bentonitic clay. *Minerals* **2016**, *6*, 129. [CrossRef]

48. Oueslati, W.; Ammar, M.; Chorfi, N. Quantitative XRD analysis of the structural changes of Ba-exchanged montmorillonite: Effect of an in situ hydrous perturbation. *Minerals* **2015**, *5*, 507–526. [CrossRef]

49. Aivalioti, M.; Vamvasakis, I.; Gidarakos, E. BTEX and MTBE adsorption onto raw and thermally modified diatomite. *J. Hazard. Mater.* **2010**, *178*, 136–143. [CrossRef] [PubMed]

50. Bastani, D.; Safekordi, A.A.; Alihosseini, A.; Taghikhani, V. Study of oil sorption by expanded perlite at 298.15 K. *Sep. Purif. Technol.* **2006**, *52*, 295–300. [CrossRef]

51. Teas, C.; Kalligeros, S.; Zanikos, F.; Stournas, S.; Lois, E.; Anastopoulos, G. Investigation of the effectiveness of absorbent materials in oil spills clean up. *Desalination* **2001**, *140*, 259–264. [CrossRef]

52. Carmody, O.; Frost, R.; Xi, Y.; Kokot, S. Adsorption of hydrocarbons on organo-clays—implications for oil spill remediation. *J. Colloid Interface Sci.* **2007**, *305*, 17–24. [CrossRef] [PubMed]

53. Simpson, J.A.; Bowman, R.S. Nonequilibrium sorption and transport of volatile petroleum hydrocarbons in surfactant-modified zeolite. *J. Contam. Hydrol.* **2009**, *108*, 1–11. [CrossRef] [PubMed]

54. Wang, D.; McLaughlin, E.; Pfeffer, R.; Lin, Y.S. Adsorption of oils from pure liquid and oil–water emulsion on hydrophobic silica aerogels. *Sep. Purif. Technol.* **2012**, *99*, 28–35. [CrossRef]

55. Emam, E.A. Modified activated carbon and bentonite used to adsorb petroleum hydrocarbons emulsified in aqueous solution. *Am. J. Environ. Prot.* **2013**, *2*, 161–169. [CrossRef]

56. Sakthivel, T.; Reid, D.L.; Goldstein, I.; Hench, L.; Seal, S. Hydrophobic high surface area zeolites derived from fly ash for oil spill remediation. *Environ. Sci. Technol.* **2013**, *47*, 5843–5850. [CrossRef] [PubMed]

57. Vidal, C.B.; Raulino, G.S.C.; Barros, A.L.; Lima, A.C.A.; Ribeiro, J.P.; Pires, M.J.R.; Nascimento, R.F. BTEX removal from aqueous solutions by HDTMA-modified Y zeolite. *J. Environ. Manag.* **2012**, *112*, 178–185. [CrossRef] [PubMed]

58. Szala, B.; Bajda, T.; Matusik, J.; Zięba, K.; Kijak, B. BTX sorption on Na-P1 organo-zeolite as a process controlled by the amount of adsorbed HDTMA. *Microporous Mesoporous Mater.* **2015**, *202*, 115–123. [CrossRef]

59. Muir, B.; Matusik, J.; Bajda, T. New insights into alkylammonium-functionalized clinoptilolite and Na-P1 zeolite: Structural and textural features. *Appl. Surf. Sci.* **2016**, *361*, 242–250. [CrossRef]

60. Muir, B.; Bajda, T. Organically modified zeolites in petroleum compounds spill cleanup—Production, efficiency, utilization. *Fuel Process. Technol.* **2016**, *149*, 153–162. [CrossRef]

61. Moazed, H.; Viraraghavan, T. Removal of oil from water by bentonite organoclay. *Pract. Period. Hazard. Toxic Radioact. Waste Manag.* **2005**, *9*, 130–134. [CrossRef]

62. Alther, G.R. Organically modified clay removes oil from water. *Waste Manag.* **1995**, *15*, 623–628. [CrossRef]

63. Mowla, D.; Karimi, G.; Salehi, K. Modeling of the adsorption breakthrough behaviors of oil from salty waters in a fixed bed of commercial organoclay/sand mixture. *Chem. Eng. J.* **2013**, *218*, 116–125. [CrossRef]

64. Sarkar, B.; Megharaj, M.; Shanmuganathan, D.; Naidu, R. Toxicity of organoclays to microbial processes and earthworm survival in soils. *J. Hazard. Mater.* **2013**, *261*, 793–800. [CrossRef] [PubMed]

65. Michel, M.M. Sorpcja oleju na złożach mineralnych. The sorption of oil on the mineral beds. *Sci. Rev. Eng. Environ. Sci.* **2005**, *1*, 95–102. (In Polish)

66. Bandura, L.; Franus, M.; Panek, R.; Woszuk, A.; Franus, W. Characterization of zeolites and their use as adsorbents of petroleum substances. *Przem. Chem.* **2015**, *94*, 323–327.

67. *Standard Test Method for Sorbent Performance of Adsorbents*; ASTM F726-99; American Society for Testing and Materials: West Conshohocken, PA, USA, 1999.

68. Zhao, M.Q.; Huang, J.Q.; Zhang, Q.; Luo, W.L.; Wei, F. Improvement of oil adsorption performance by a sponge-like natural vermiculite-carbon nanotube hybrid. *Appl. Clay Sci.* **2011**, *53*, 1–7. [CrossRef]

69. Ankowski, A. Wykorzystanie zeolitów z popiołów lotnych w sorpcji substancji ropopochodnych w warunkach rzeczywistych. In Proceedings of the Debata o Przyszłości Energetyki, Wysowa Zdrój, Poland, 4–7 May 2010. (In Polish)

70. Zadaka-Amir, D.; Bleiman, N.; Mishael, Y.G. Sepiolite as an effective natural porous adsorbent for surface oil-spill. *Microporous Mesoporous Mater.* **2013**, *169*, 153–159. [CrossRef]

71. Bandura, L.; Franus, M.; Józefaciuk, G.; Franus, W. Synthetic zeolites from fly ash as effective mineral sorbents for land-based petroleum spills cleanup. *Fuel* **2015**, *147*, 100–107. [CrossRef]

72. Franus, M.; Jozefaciuk, G.; Bandura, L.; Lamorski, K.; Hajnos, M.; Franus, W. Modification of lightweight aggregates' microstructure by used motor oil addition. *Materials* **2016**, *9*, 845. [CrossRef]

73. Franus, W.; Jozefaciuk, G.; Bandura, L.; Franus, M. Use of spent zeolite sorbents for the preparation of lightweight aggregates differing in microstructure. *Minerals* **2017**, *7*, 25. [CrossRef]

74. Boglaienko, D.; Tansel, B. Instantaneous stabilization of floating oils by surface application of natural granular materials (beach sand and limestone). *Mar. Pollut. Bull.* **2015**, *91*, 107–112. [CrossRef] [PubMed]

75. Boglaienko, D.; Tansel, B. Encapsulation of light hydrophobic liquids with fine quartz sand: Property based characterization and stability in aqueous media with different salinities. *Chem. Eng. Sci.* **2016**, *145*, 90–96. [CrossRef]

76. Boglaienko, D.; Tansel, B.; Sukop, M.C. Granular encapsulation of light hydrophobic liquids (LHL) in LHL-salt water systems: Particle induced densification with quartz sand. *Chemosphere* **2016**, *144*, 1358–1364. [CrossRef] [PubMed]

77. Boglaienko, D.; Tansel, B. Gravity induced densification of floating crude oil by granular materials: Effect of particle size and surface morphology. *Sci. Total Environ.* **2016**, *556*, 146–153. [CrossRef] [PubMed]

78. Carmody, O.; Frost, R.; Xi, Y.; Kokot, S. Surface characterisation of selected sorbent materials for common hydrocarbon fuels. *Surf. Sci.* **2007**, *601*, 2066–2076. [CrossRef]

79. Carmody, O.; Frost, R.; Xi, Y.; Kokot, S. Selected adsorbent materials for oil-spill cleanup: A thermoanalytical study. *J. Therm. Anal. Calorim.* **2008**, *91*, 809–816. [CrossRef]

80. Carvalho, M.N.; da Motta, M.; Benachour, M.; Sales, D.C.S.; Abreu, C.A.M. Evaluation of BTEX and phenol removal from aqueous solution by multi-solute adsorption onto smectite organoclay. *J. Hazard. Mater.* **2012**, *239–240*, 95–101. [CrossRef] [PubMed]

81. Nourmoradi, H.; Nikaeen, M.; Khiadani, M. Removal of benzene, toluene, ethylbenzene and xylene (BTEX) from aqueous solutions by montmorillonite modified with nonionic surfactant: Equilibrium, kinetic and thermodynamic study. *Chem. Eng. J.* **2012**, *191*, 341–348. [CrossRef]

82. Aivalioti, M.; Pothoulaki, D.; Papoulias, P.; Gidarakos, E. Removal of BTEX, MTBE and TAME from aqueous solutions by adsorption onto raw and thermally treated lignite. *J. Hazard. Mater.* **2012**, *207–208*, 136–146. [CrossRef] [PubMed]

83. Standeker, S.; Novak, Z.; Knez, Z. Removal of BTEX vapours from waste gas streams using silica aerogels of different hydrophobicity. *J. Hazard. Mater.* **2009**, *165*, 1114–1118. [CrossRef] [PubMed]

84. Sharmasarkar, S.; Jaynes, W.F.; Vance, G.F. BTEX sorption by montmorillonite organo-clays: TMPA, ADAM, HDTMA. *Water Air Soil Pollut.* **2000**, *119*, 257–273. [CrossRef]

85. Simantiraki, F.; Aivalioti, M.; Gidarakos, E. Laboratory study on the remediation of BTEX contaminated groundwater using compost and Greek natural zeolite. In Proceedings of the CRETE 2012 3rd International Conference on Industrial and Hazardous Waste Management, Chania, Greece, 12–14 September 2012; pp. 1–8.

86. Torabian, A.; Kazemian, H.; Seifi, L.; Bidhendi, G.N.; Azimi, A.A.; Ghadiri, S.K. Removal of petroleum aromatic hydrocarbons by surfactant-modified natural zeolite: The effect of surfactant. *Clean Soil Air Water* **2010**, *38*, 77–83. [CrossRef]

87. Seifi, L.; Torabian, A.; Kazemian, H.; Bidhendi, G.N.; Azimi, A.A.; Farhadi, F.; Nazmara, S. Kinetic study of BTEX removal using granulated surfactant-modified natural zeolites nanoparticles. *Water Air Soil Pollut.* **2011**, *219*, 443–457. [CrossRef]

88. Ho, Y.; McKay, G. Pseudo-second order model for sorption processes. *Process Biochem.* **1999**, *34*, 451–465. [CrossRef]

89. Long, C.; Li, Y.; Yu, W.; Li, A. Removal of benzene and methyl ethyl ketone vapor: Comparison of hypercrosslinked polymeric adsorbent with activated carbon. *J. Hazard. Mater.* **2012**, *203–204*, 251–256. [CrossRef] [PubMed]

90. Alejandro, S.; Valdés, H.; Manéro, M.H.; Zaror, C.A. Oxidative regeneration of toluene-saturated natural zeolite by gaseous ozone: the influence of zeolite chemical surface characteristics. *J. Hazard. Mater.* **2014**, *274*, 212–220. [CrossRef] [PubMed]

91. Valdés, H.; Solar, V.A.; Cabrera, E.H.; Veloso, A.F.; Zaror, C.A. Control of released volatile organic compounds from industrial facilities using natural and acid-treated mordenites: The role of acidic surface sites on the adsorption mechanism. *Chem. Eng. J.* **2014**, *244*, 117–127. [CrossRef]

92. Cosseron, A.F.; Daou, T.J.; Tzanis, L.; Nouali, H.; Deroche, I.; Coasne, B.; Tchamber, V. Adsorption of volatile organic compounds in pure silica CHA, BEA, MFI and STT-type zeolites. *Microporous Mesoporous Mater.* **2013**, *173*, 147–154. [CrossRef]

93. Kim, K.J.; Ahn, H.G. The effect of pore structure of zeolite on the adsorption of VOCs and their desorption properties by microwave heating. *Microporous Mesoporous Mater.* **2012**, *152*, 78–83. [CrossRef]

94. Serrano, D.P.; Calleja, G.; Botas, J.A.; Gutierrez, F.J. Characterization of adsorptive and hydrophobic properties of silicalite-1, ZSM-5, TS-1 and Beta zeolites by TPD techniques. *Sep. Purif. Technol.* **2007**, *54*, 1–9. [CrossRef]

95. Bandura, L.; Panek, R.; Rotko, M.; Franus, W. Synthetic zeolites from fly ash for an effective trapping of BTX in gas stream. *Microporous Mesoporous Mater.* **2016**, *223*, 1–9. [CrossRef]

minerals

MDPI

Article

Experimental Study of Montmorillonite Structure and Transformation of Its Properties under Treatment with Inorganic Acid Solutions

Victoria V. Krupskaya [1,2,*], Sergey V. Zakusin [1,2], Ekaterina A. Tyupina [3,4], Olga V. Dorzhieva [1,5], Anatoliy P. Zhukhlistov [1], Petr E. Belousov [1] and Maria N. Timofeeva [6]

[1] Belov Laboratory of Mineral Cristallochemistry, Institute of Ore Deposits, Petrography, Mineralogy and Geochemistry, Russian Academy of Science, Moscow 119017, Russia; zakusinsergey@gmail.com (S.V.Z.); dorzhievaov@gmail.com (O.V.D.); anzhu@igem.ru (A.P.Z.); pitbl@mail.ru (P.E.B.)
[2] Department of Ecological Geology, Faculty of Geology, Lomonosov Moscow State University, Moscow 119899, Russia
[3] Institute of Modern Energetics and Nanotechnology Materials-IPC, Dmitry Mendeleev University of Chemical Technology of Russia, Moscow 125480, Russia; tk1972@mail.ru
[4] National Research Nuclear University, Moscow 115409, Russia
[5] Laboratory of Physical Methods Application to the Study of Rock-forming Minerals, Geological Institute, Russian Academy of Sciences, Moscow 119017, Russia
[6] Research Group for Heterogeneous Catalysts for Liquid-Phase Selective Oxidations, Boreskov Institute of Catalysis SB RAS, Novosibirsk 630090, Russia; timofeeva@catalysis.ru
* Correspondence: krupskaya@ruclay.com; Tel.: +7-499-230-82-96

Academic Editor: Annalisa Martucci
Received: 17 December 2016; Accepted: 19 March 2017; Published: 23 March 2017

Abstract: This paper discusses the mechanism of montmorillonite structural alteration and modification of bentonites' properties (based on samples from clay deposits Taganskoye, Kazakhstan and Mukhortala, Buriatia) under thermochemical treatment (treatment with inorganic acid solutions at different temperatures, concentrations and reaction times). Treatment conditions were chosen according to those accepted in chemical industry for obtaining acid modified clays as catalysts or sorbents. Also, more intense treatment was carried out to simulate possible influence at the liquid radioactive site repositories. A series of methods was used: XRD, FTIR, ICP-AES, TEM, nitrogen adsorption, and particle size analysis. It allowed revealing certain processes: transformation of montmorillonite structure which appears in the leaching of interlayer and octahedral cations and protonation of the interlayer and –OH groups at octahedral sheets. In turn, changes in the structure of the 2:1 layer of montmorillonite and its interlayer result in significant alterations in the properties: reduction of cation exchange capacity and an increase of specific surface area. Acid treatment also leads to a redistribution of particle sizes and changes the pore system. The results of the work showed that bentonite clays retain a significant portion of their adsorption properties even after a prolonged and intense thermochemical treatment (1 M HNO_3, 60 °C, 108 h).

Keywords: engineered barriers; bentonite clays; thermochemical treatments; montmorillonite; structural modification; adsorption properties

1. Introduction

Clay minerals are widely used in various industries including radioactive waste management as a component of barrier systems for waste disposal. These systems are used for radioactive waste repositories of different levels of activity and are intended to provide safe storage for several hundreds or thousands years due to their high adsorption capacity for radionuclides and low water permeability [1–3].

Bentonite buffer between containers and the tunnel walls achieves several strategic outcomes at once: it provides access restriction of groundwater to the radioactive waste (RW) creating the conditions under which mass transfer between the waste and underground waters is only possible by diffusion; it suppresses migration of the radionuclides in colloidal form into the groundwater; it ensures effective sorption of radionuclides after possible depressurization of radioactive waste containers; it seals open cracks and large pores in the rocks due to the high swelling capacity, etc. [2].

One of the main components of the engineered barriers is bentonite, which contains 70–95% montmorillonite, dioctahedral specie of the smectite mineral group. Montmorillonite is a 2:1 type hydrous aluminosilicate with the octahedral sheet "sandwiched" between two tetrahedral sheets. Cation substitution in tetrahedral and mostly in octahedral sites provides a negative layer charge of about 0.2–0.5 eV. Layer charge is compensated by the introduction of exchangeable interlayer cations (Na^+, Ca^{2+}, Mg^{2+}, etc.) usually in hydrated form [4–8] which in turn provides adsorption sites on the inner and outer surface of the crystal. These particularities of montmorillonite structure determine specific properties of bentonite clays, especially high adsorption capacity towards heavy metals such as cesium, plutonium etc., which are commonly found in radioactive wastes.

Construction concept of waste disposal in the Russian Federation also includes the use of bentonite clay as a component of an engineered multi-barrier system [9,10]. According to current concepts, a bentonite barrier must retain its properties for over thousands of years. Thus, when analyzing the prospects of using bentonite clays, one needs to consider not only their sorption properties in their natural state, but a possible loss of sorption capacity and other parameters needed to preserve the stability of a bentonite barrier in aggressive environments. The most aggressive environment for bentonite clays is acid solutions. A number of processes take place under acid treatment and lead to a significant transformation of the structure and properties of montmorillonite bentonite clays [11–14]. A significant increase in the specific surface of the acid-modified clays promotes wide use of such material for removal of heavy metals, radionuclides and for oil refining [15–19]. The transformation of the structure and properties of bentonite under the influence of acids, mostly under sulfuric and hydrochloric acids, have been studied by different authors [20–26]. Using ultrasound and microwaves in addition to an acid treatment intensifies smectite mineral transformation [27]. In industry, acid-modified bentonites are most commonly used as catalysts [28–32].

In Russia, studies have been conducted to identify the most thermochemically resistant bentonite in order to determine the most appropriate bentonite clay for engineered barriers for waste disposal. The increase of the temperature in the vicinity of radioactive waste occurs due to radioactive decay. Thermochemical effects were modeled based on inorganic acid solutions (nitric and hydrochloric) at elevated temperatures. Due to its toxicity, nitric acid is not used in the chemical industry for the production of modified materials. Currently in Russia, there are plants that still dump liquid radioactive waste into deep layers of geological structures and in artificial surface repositories [33–37]. Nitric acid is used to prepare liquid radioactive waste for disposal. These concepts are obsolete and are being withdrawn from use. Bentonite is also used to bury surface basins, such as Lake Karachai [38].

However, in the long run, bentonite clays may be affected by aggressive fluids that are most likely to decrease insulating properties of the barrier. The aim of this study was to evaluate the mechanism of montmorillonite transformation under the acid solution treatment as well as its influence on bentonite properties.

2. Materials and Methods

Fine clay fractions (<1 μm) separated from bentonites from Mukhortala (Buriatia, sample Mt-M) and Taganskoye (Kazakhstan, sample Mt-T1 and Mt-T2 (numbers "1" and "2" correspond to different treatments as will be shown below)) deposits were used for this study. Montmorillonite content in Mt-T and Mt-M sample were 70% and 97%, respectively. Impurities in Mt-M sample were presented by

20–30% opal C/A, which could not be removed with the sedimentation technique, and quartz (about 2–3%) in Mt-T sample.

All experiments on acid treatment were carried out in closed systems under static conditions. For this purpose, bentonite samples were placed in sealed vessels with acid solution in a weight ratio of 1:100. Vessels were in turn placed in an oven and aged for the desired time. The experiments with HCl solutions (0.125, 0.25, 0.5, 1, 3 M) were carried out at room temperature for 7 days (Mt-T1 and Mt-M samples) and with 1 M HNO_3 solutions (Mt-T2 sample) at 60 °C for 12, 36, 50, and 108 h. The bentonite samples from the Taganskoye deposit have quite a different composition and were named Mt-T1 for the experiment with HCl solution and Mt-T2 for the experiment with HNO_3 solution. Hydrochloric acid treatment was conducted under the conditions accepted in chemical industry for modified bentonite production. Experiments with nitric acid treatment were conducted at high temperatures in order to simulate the conditions that may occur during liquid radioactive waste disposal.

Initial and modified samples were analyzed with a series of methods. X-ray diffraction patterns were obtained with X-ray diffractometer Ultima-IV (Rigaku, Tokyo, Japan) acquired with the funding of Moscow State University Development Program (Cu-Kα radiation, semiconductor 1D detector D/Tex-Ultra, scan range 3–65° (2θ), scan speed 5°/min and step—0.02° (2θ)). Partially oriented specimens were prepared by pressing powder into a sample holder.

The chemical analyses of the solids were carried out by means of inductively coupled plasma atomic emission spectrometry (ICP-AES) using ICPE-9000 equipment (Shimadzu, Kyoto, Japan).

Excessive pressure and surface leveling leads to partial orientation of the montmorillonite particles in the specimen plane. We did not succeed in preparing samples with a good degree of particle orientation from drops of aqueous suspensions because some samples had been significantly modified after the treatment so they could not form thin films on a glass surface. The results were analyzed according to Drits, Kossovskaya [4], Moore and Reynolds [5]. Mineral composition was estimated using the Rietveld method [39] with PROFEX GUI for BGMN [40].

Fourier transform infrared spectroscopy analysis was carried out using FTIR spectrometer Vertex 80v equipped by DTGS detector and KBr beam-splitter (Bruker, Ettlingen, Germany). The adsorption spectra recordings were performed in vacuum in the 4000–400 cm^{-1} wavelength range with 256 scans for each sample and the resolution of 4 cm^{-1}. Samples were prepared as pressed KBr-pellets: 0.5 mg of sample was dispersed in 200 mg of KBr; this mixture was placed in a 13 mm pellet die and pressed in vacuum for 1 h. Spectra manipulations were performed using the OPUS 7.1 software (Bruker, Ettlingen, Germany). Baseline correction was made automatically by Concave Rubberband method with 64 baseline points and 10 iterations.

Natural montmorillonites (Mt-M, Mt-T1, Mt-T2) as well as some samples after treatment were selected based on the results of X-ray diffraction studies (Mt-M—3.0 M HCl; Mt-T1—0.25 M and 0.5 M HCl; 12 and 108 h) for further investigation by high-resolution transmission electron microscopy (HRTEM) using JEM-2100 with an X-Max attachment for X-ray energy dispersive analysis.

Evaluation of a specific surface area was carried out using the Analyzer Quadrasorb SI/Kr (Quantachrome Instruments, Boynton Beach, FL, USA). Adsorption was performed at the liquid nitrogen temperature (77.35 K). Nitrogen with a 99.999% purity served as an adsorbate. Helium grade 6.0 (99.9999%) was used for the volume calibration of the measuring cells. Calculation was carried out by the BET multiple-point isotherm in the range of P/P0 from 0.05 to 0.30. Samples were pre-dried in vacuum at 100 °C.

Cation exchange capacity was determined by triethylenetetramine copper complex $[Cu(Trien)]^{2+}$ adsorption method [41].

Particle size distribution was evaluated with the laser diffraction technique by Fritsch ANALYSETTE 22 NanoTec (Fritsch, Idar-Oberstein, Germany) equipped with 70 W and 36 kHz ultrasonic emitter in the size range 0.01–1000 µm Data processing was carried out by algorithm based on Fredholm integral equations.

3. Results and Discussion

3.1. Transformation of the Montmorillonite Structure under Acid Treatment

Considerable crystal-chemical transformations—in particular, changes of montmorillonite micromorphology and adsorption characteristics—were detected under treatment with hydrochloric and nitric acid solutions. A variety of processes that modified structure and properties of montmorillonite particles were observed during the treatment with inorganic acid solutions: dissolution of carbonates and feldspars, destruction of the most defective phyllosilicate phases (e.g., nanosized smectites), removal of cations from the interlayer spacing, substitution of interlayer cations with oxonium ion, leaching of the octahedral cations, and finally, the complete destruction of the structure.

Studied montmorillonite samples are characterized by the heterogeneous composition of the interlayer: Ca^{2+} and Mg^{2+} for the Mt-M sample (d_{001} = 15.3 Å), and Ca^{2+}, Mg^{2+} and Na^+ (d_{001} = 13.7 Å and d_{001} = 13.9 Å, respectively) for Mt-T1 and Mt-T2 samples (Figure 1, Table 1). Presence of Na^+ in the montmorillonites from the Taganskoye deposit was also confirmed in previous studies [26,42,43].

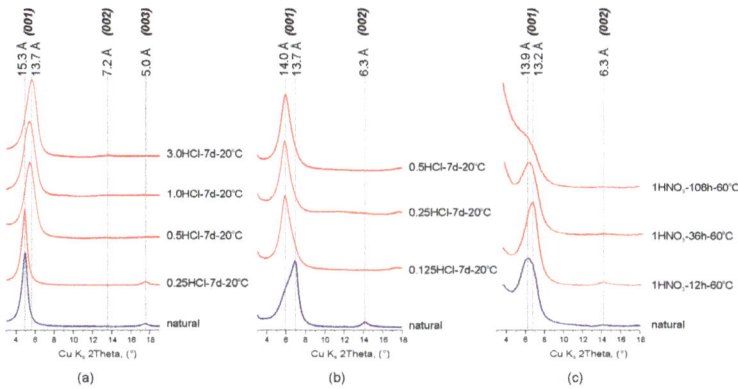

Figure 1. Fragments of the X-ray diffraction patterns of natural and treated montmorillonites: (**a**) Mt-M; (**b**,**c**) Mt-T. Abbreviations: 3.0 HCl, 1 HNO_3, type and concentration (given in mol/L) of acid; 7 d (days), 12 h (hours), time of treatment, 20, 90 °C, temperature of treatment.

Table 1. Changes of the structural parameters, textural and adsorption characteristics of montmorillonites due to acid activation (Specific Surface Area: S_{BET}, Total Pore Value: V_Σ, Cation Exchange Capacity: CEC).

Sample	Treatment			Interlayer Space d_{001} (Å)	Particle Thickness h(00l) (CSR) (nm)	Number of Layers (N)	S_{BET} (m²/g)	V_Σ (cm³/g)	CEC (meq/100 g)
	Acid (mol/L)	T (°C)	Time Days/Hours						
					HCl				
	-	-	-	15.3	10.7	7	77	0.211	46
	0.25	20	7 d	15.3	12.2	8	79	0.219	43
Mt-M	0.5	20	7 d	14.0	7.0	5	92	0.243	-
	1.0	20	7 d	14.0	5.6	4	95	0.256	49
	3.0	20	7 d	13.7	5.5	4	101	0.328	64
	-	-	-	13.7	9.6	7	42	0.074	75
Mt-T1	0.125	20	7 d	14.0	7.0	5	45	0.084	65
	0.25	20	7 d	14.0	7.0	5	50	0.095	64
	0.5	20	7 d	14.0	5.6	4	51	0.083	70
					HNO_3				
	-	-	-	13.9	8.2	6	67	0.085	86
Mt-T2	1	60	12 h	14.0	7.0	5	110	0.114	58
	1	60	36 h	14.0	7.0	5	191	0.192	56
	1	60	108 h	13.8	5.5	4	301	0.353	40

Treatment of Mt-M sample with hydrochloric acid solution with various concentrations leads to partial protonation of the interlayer which is observed by a decrease in d_{001} values and changes in basal reflections series from d_{001} = 15.3 Å, d_{003} = 5.0 Å, d_{005} = 3.0 Å to d_{001} = 13.7 Å. d_{002} = 7.2 Å, d_{003} = 4.8 Å, d_{004} = 3.6 Å (Figure 1). These changes can be related with leaching of the interlayer cations (Ca^{2+} and Mg^{2+}) and partial protonation of interlayer space i.e., substitution of interlayer cations with oxonium ion [44,45] because protonation of Ca–Mg montmorillonite proceeds much faster than that of Na-forms [43].

Early, we demonstrated that the chemical composition and structural characteristics of Mt depended on the HNO_3 and HCl concentration [26,32]. Changes of chemical composition and interlayer space were negligible after modification of montmorillonite under the concentrations up to 0.5 M. These changes were noticeable only after treatment with 3.0 M acid. Here we investigated effect of exposure time of 1M HNO_3 at 60 °C on chemical composition of Mt-T2. The main results are shown in the Table 2. Experimental data point that leaching of interlayer cation is observed after the treatment with 1 M HNO_3 for 12 h.

Table 2. Chemical composition of MM-T2 natural and modified by 1 M HNO_3 at 60 °C.

Time (h)	Chemical Composition (wt %)						
	Si	Al	Fe	Mg	Ca	Na	Si/Al
-	26.4	7.8	3.9	1.9	0.6	2.4	3.4
12	27.3	7.6	3.7	1.6	trace	trace	3.6
36	28.0	6.9	3.2	1.4	0.2	trace	4.1
108	31.8	5.0	2.0	0.9	trace	trace	6.3

Profile alignment of basal reflections (001) and displacement of maximum 13.7 Å to 14.0 Å was observed after treatment of Mt-T1 sample with hydrochloric acid solution. Irregular profile shape of d_{001} reflection indicates the presence of two possible montmorillonite phases with different interlayer cations in the sample. This was also found by other researchers of the Taganskoye montmorillonite deposit [43]. Since any other impurities except quartz were not found in this sample, the described change in the XRD patterns can be attributed to the partial destruction (dissolution) of the most defective and probably nanosized particles of Na-montmorillonite.

Treatment of Mt-T sample with 1M HNO_3 solution for 12 h leads to a displacement of the d_{001} reflection from 13.9 to 13.2 Å with an increase of its intensity. This fact is explained by the dissolution of the defective part of the montmorillonite and enhancement of the interlayer ordering by partial substitution of Ca^{2+} and Mg^{2+} cations. Further treatment of the sample with nitric acid solution reduces the layer stacking ordering degree and leads to disintegration of montmorillonite 2:1 layer structure. In XRD patterns, the described process is manifested by a decrease in the peaks intensity and broadening of the basal reflections until the complete disappearance except (001) reflection. Displacement of the (001) reflection back to 13.9 Å after the 36 and 108 h of treatment can be explained by the Mg and Al octahedral cations leaching and their migration into the interlayer space of montmorillonite.

The changes in the sizes of the coherent scattering domains (CSD) along the c-axis were calculated in accordance with the Scherrer equation for the (001) reflection of natural and modified montmorillonites (Table 1). In general, the size of CSD along the c-axis corresponds with the crystallite thickness [46]. The average crystallite thicknesses of the samples equals 10.7 nm which corresponds to 7 layers (7 N), 9.6 nm (7 N), 8.2 nm (6 N) for natural Mt-M, MtM-T1 and Mt-T2 samples respectively, and to 5.5–5.6 nm (4 N) for the acid treated montmorillonites. A decrease of CSD sizes is related to a consequent disintegration of montmorillonite particles and increase of the stacking faults quantity.

The increase of the adsorbed water bands on IR spectra in the range of 4000–2500 cm^{-1} probably indicates a decrease of crystallite size (Figure 2a) after the treatment of Mt-T2 sample with HNO_3 acid for 108 h. Previous studies of Mt-M and Mt-T1 samples treated with HCl acid [26] showed the reduction of band intensity at 841–845 cm^{-1} (Al–Mg–OH), 882 cm^{-1} (Al–Fe–OH) and 914–926 cm^{-1} (Al–Al–OH), which indicates the leaching of Al^{3+}, Mg^{2+} and Fe^{3+} cations from the octahedral sites [44–49]. Reduction of the band intensities at 925 and 876 cm^{-1} is shown at Figure 2c. It corresponds to Al–Al–OH and Al–Fe–OH vibrations in the structure of octahedral sheet of montmorillonite (Mt-T2 sample) during the treatment with the nitric acid. It may indicate a leaching of Fe^{3+} and Al^{3+} from the octahedral sites.

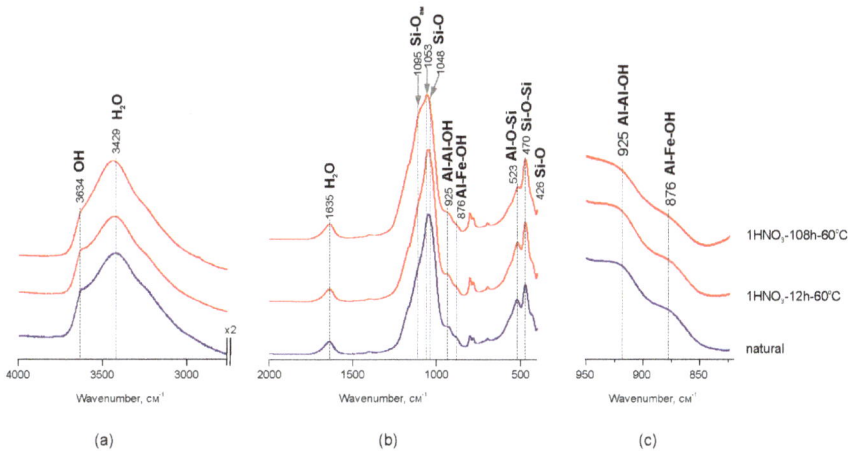

Figure 2. Areas of interest on MM-T IR spectra before and after treatment with 1 M HNO$_3$ 60 °C: (**a**,**b**) the characteristic part of spectrum; (**c**) a magnified part of the spectra.

Leaching of the octahedral cations leads to modifications of the interaction between the octahedral and tetrahedral sheets in the 2:1 layer and, as a result, to the partial destruction of octahedral sheets. Alterations in FTIR data revealed as decreasing intensities of the bands related to octahedral Al, Fe and Mg and exchangeable Ca, Na and Mg correlate with changes in chemical composition shown above. Therefore, the structural units comprised of the tetrahedral sheet fragments are released from the 2:1 structure, which corresponds to "Si–O$_{free}$" vibrations—1095 cm^{-1} [11,15,21,50]. Changes in the Si–O–Si band profile within the range of 1100–1050 cm^{-1} (Figure 2b) indicate transformation of interaction within the tetrahedral sheet. In particular, displacement of the maximum in the natural sample from 1050 to 1095 cm^{-1} after treating it with 1 M HNO$_3$ for 108 h indicates the increase of amorphous silica content due to partial destruction of the tetrahedral sheet which is confirmed by chemical analysis data as an increase in Si content in powder samples.

Transmission electron microscopy (TEM) allows collection of data at the level of crystal lattice and helps to estimate structural transformation of montmorillonite, properties of surface and interactions between the layers during its treatment with inorganic acid solutions, and changes in montmorillonite particles' chemical composition [51]. In natural samples from the Tagansoye deposit (Figure 3a) among the laminar montmorillonite particles with a size of 1–2 μm, there is a significant amount of small and thin nano-sized particles that cover the specimen and produce grey background in micrographs. Particle identification was carried out based on microdiffraction patterns and micro-area chemical analysis. The nanosized phase is characterized by a higher content of Na as an interlayer cation and Fe is located in the octahedral sheet due to the isomorphic substitution of Al. Under hydrochloric acid with lower concentrations, as well as under the treatment with nitric acid for shorter periods, almost full dissolution of nano-sized particles as the less acid resistant phase takes place and the grey background on micrographs is not observed. Besides, these laminar particles have a lot of folded edges which are not usually observed in acid-modified montmorillonites. Bentonites from Mukhortala deposit (Mt-M) contain a lot of opal distributed in pores between thin and relatively thick laminar montmorillonite particles and also on their relatively clean surfaces (Figure 3c).

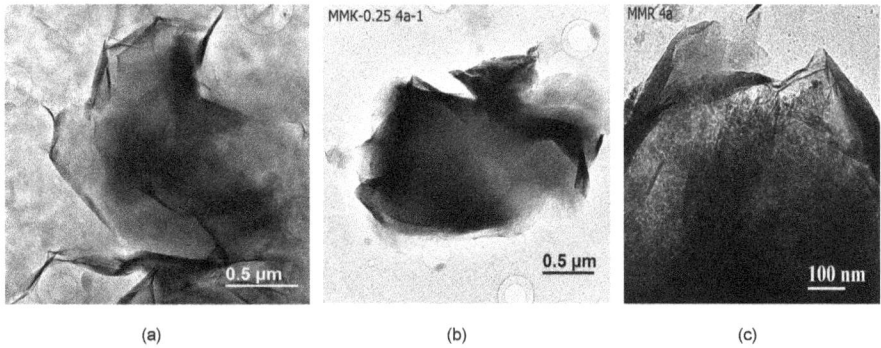

Figure 3. Transmission electron microscopic (TEM) images: (**a**) natural particles from the sample MM-T; (**b**) treated with 0.25 M HCl; (**c**) natural sample MM-M.

Studied samples of the natural montmorillonites are characterized by peculiar particle morphology that is typical for montmorillonites of different genesis and by a wide spread of welted edges, which can be useful for obtaining pictures of the lattices (Figure 4). Images obtained from edges of natural montmorillonite (Mt-T2 sample) show the lattice stripes that correspond to the basal planes (Figure 4a,b). The width of the areas with lattice stripes is about 23–32 nm. The value of the interlayer space for different particles varies within the range from 10.5 to 11.8–12.3 Å.

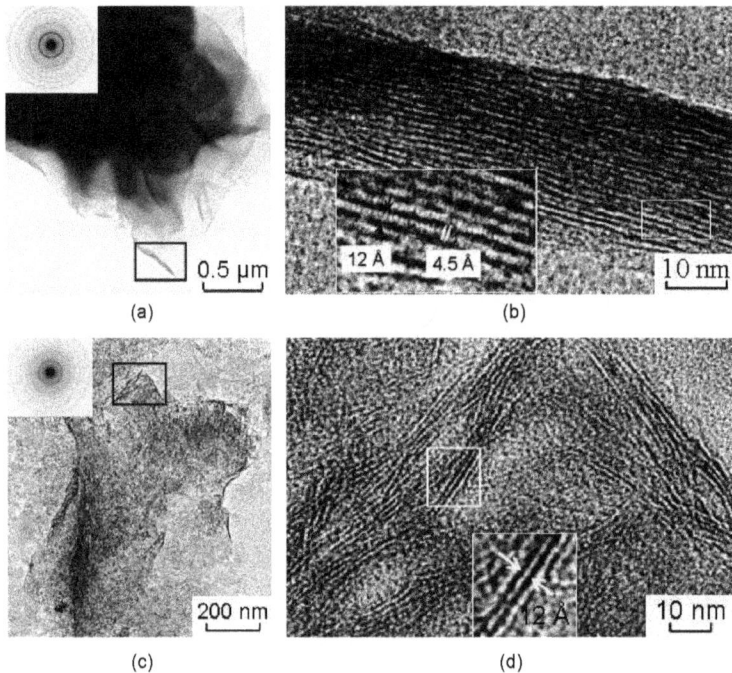

Figure 4. Transmission electron microscopic (TEM) images of natural montmorillonite particles (**a,b**), and treated with 1 M HNO_3 for 108 h (**c,d**). Microdiffraction figures are located at enlarged areas. (**b,d**) high resolution images from the folded edge of the montmorillonite particles (area locations are shown as black rectangles in Figure (**a,c**), respectively).

The lattice stripes of different particles can be pulled simultaneously, maintaining a noticeable distance between the adjacent bands or showing some curving of the surface, which indicates only a small change of interlayer space along the layers. Moreover, there are areas of the image with two-dimensional lattice stripes (Figure 4b). Basal lattice stripes are crossed by lattice stripes with an interlayer space of 4.5 Å which corresponds to planes (020) and indicates local manifestations of coherence in stacking of adjacent layers.

Microaggregate surface of Mt-T2 sample became uneven as the result of the treatment with HNO_3 for 108 h. This effect can be seen only at some fragments of the images with folded edges with typical parallel basal lattice stripes from planes (001). It correlates to the small packs of 2–3 layers (Figure 4b) less than 25 nm long. The interlayer spacing 12–14 Å indicates the presence of initial montmorillonite. The observed pattern of distribution of the basal lattice stripes reflects the nano-level changes in the surface morphology of montmorillonite. It implies an existence of thin packages of layers where the planes (001) are arranged perpendicularly to the surface of the particles. This effect in particular leads to an increase of its surface and the occurrence of active sites for the selective sorption of heavy metals and radionuclides.

3.2. Transformation of Montmorillonite Adsorption Properties under Acid Treatment

Specific surface area (BET) and cation exchange capacity (CEC) can be qualified as characteristics of adsorption properties of bentonite clays. CEC values of montmorillonites depend most significantly on the amount of isomorphic substitutions in octahedral sites. In turn, specific surface area is primarily controlled by degree of fineness, amount of impurities, particle charge, their ability to coagulate, etc. Mt-M sample contains a significant amount of opal which provides a relatively higher specific surface area than that of Taganskoye deposit samples (Mt-T1 and Mt-T2). Commonly, specific surface area (BET) values are in direct relationship with CEC value [52]. Increasing of BET value leads to an increase of CEC value and vice versa. In most cases, the above is true for natural soils with different mineral compositions [46]. However, the ratio of these indexes for montmorillonites of different composition may be in a more complex relationship. This dependence can be clearly seen in the Table 1 as an increase of specific surface area and a decrease in cation exchange capacity due to the acid treatment.

Coherent scattering domain sizes are commonly used to analyze the size of crystallites [46,53–55]. There is a cumulative index of physical-chemical activity of the most clay minerals which have correlations with specific surface area and adsorption characteristics. Thus, reduction of crystallite size should lead to an increase of specific surface area, which is shown in the study (Table 1).

At the same time, an increase of the specific surface area appears mostly in the samples which have the greatest structural changes. Thus, in the montmorillonite sample from Taganskoye deposit, treated with HNO_3 for a long period of time, the increase of the specific surface area was 301 m^2/g compared to 67 m^2/g in the natural sample. Also, considerable changes were observed in the sample with montmorillonite from the Mukhortala deposit (from 77 to 101 m^2/g, respectively).

The average pore size in all samples is approximately 5 nm and it remains constant during the experiments, while the total pore volume has been changed slightly for Mt-M and Mt-T1 samples and considerably so for Mt-T2 sample. This fact cannot be related only to the size decrease. In this way, the observed increase of specific surface area, as well as the total pore volume is related to the formation of porous structure due to the modification of montmorillonites under the acid treatment.

As a result of the treatment, a regular change is observed in the chemical composition of montmorillonite (Table 2). First of all, the reduction of Ca, Mg, Fe, and Al content and the increase of Si content is found in all samples after dissolving the nano-sized phases of montmorillonite from Taganskoye deposit. The most significant changes were found in the Mt-T2 sample treated with HNO_3 acid solution for 108 h. The increase of Si content in montmorillonite samples subjected to the intensive acid treatments connected with subsidence of amorphous silica formed during the destruction of tetrahedral sheets of 2:1 layers on the particle surface and its accumulation in the micropores.

Protonation of Al bond pairs in the octahedron transforms it from 4-coordination to 6-coordination [23] and leads to the modification of octahedral and tetrahedral sheets and to an appearance of micropores simultaneously decreasing the number of octahedral cations. Changes in the layer stacking and particles' micromorphology results in mesopore formation [14,45]. Structural transformations are shown simplified in Figure 5. Thus, there is an increase of specific surface area and pore space both capable of large cation sorption, e.g., Cs.

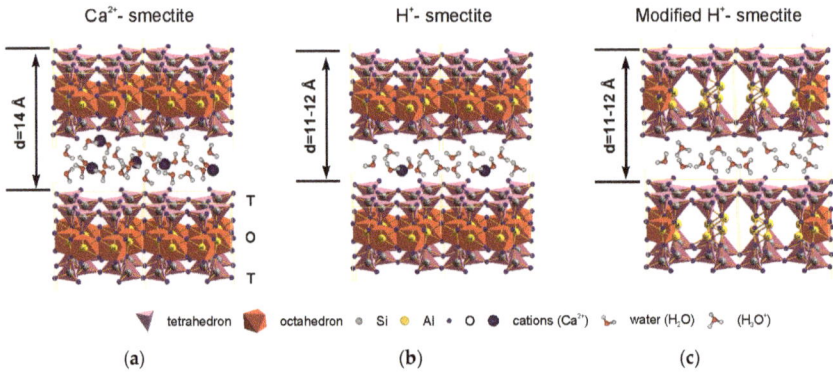

Figure 5. Schematic picture representing the structural changes in natural montmorillonite under the treatment with inorganic acid solutions: (**a**) natural Ca-montmorillonite; (**b**) partial protonation of interlayer space; (**c**) full protonation of interlayer space (H-smectite) and protonation of the OH-groups of the octahedral sheet and Al coordination change.

Critical values for describing the concepts of "micropores", "mesopores", and "macropores" vary in classifications developed for different science approaches. In engineering geology, for example, mesopore sizes are in the range of 10 to 1000 µm and micropores are from 0.1 to 10 µm [56]. V. Osipov and V. Sokolov in their morphometric investigations of soil microstructures with different composition [52] detailed the previously proposed classification of pore sizes. They divided micropores into thin (0.1–10 µm), small (1–10 µm), and large (10–100 µm) ones, suggesting the lower limit for macropores as 100 µm.

There are other classifications to consider. For example, the International Union for Pure and Applied Chemistry (the IUPAC) recommended to distinguish pores size into macropores (50 nm), mesopores (2 to 50 nm), and micropores (up to 2 nm) [57–59]. The micropores are conventionally divided into thin ultramicropores (less than 0.7 nm) and supermicropores that have an intermediate size between ultramicropore and mesopores.

In fact, all the pores in montmorillonite referred above of the size not larger than 6–9 nm for mesopores and 1–3 nm for micropores are predominantly involved in the adsorption [60]. The average pore size measured in natural and modified montmorillonites is about 5 nm which corresponds to interparticle pores according to Osipov and Sokolov [52]. At the same time, acid treatment, as shown above, leads to the appearance of pores in the structure of the layer (micropores) by partial leaching of octahedral cations, protonation of OH-groups, and changes of Al^{3+} coordination. The above process does not lead to an increase of an average pore size, however, it results in an increase of the total pore volume (Table 1).

Mesopores of a size about 5 nm and a small amount of micro-pores with a diameter of about 3 nm are predominant in the natural montmorillonite. After treatment with hydrochloric acid with concentration up to a 3.0 M at room temperature for 7 days, significant changes in the structure of the pore space were not observed, which also can be seen below in the N_2 adsorption-desorption isotherms. Treatment with 1 M HNO_3 for 12 h does not lead to significant changes of the pore space either. The most significant changes were observed as a result of treatment for 108 h (Figure 6a): the amount of 5–5.5 nm mesopores significantly increases and micropores with an average diameter of 3.5 nm appear which is not observed in natural montmorillonite samples. The particle size distribution changes even at low exposure times in the experiment with 1 M HNO_3 (Figure 6b) from monomodal shape with a maximum at 50 µm to a multimodal shape with peaks at ~80, 100, 145, and 200 µm. Thus, with increased micro- and mesoporosity, particle aggregation occurs that leads to the formation of relatively large aggregates of different sizes.

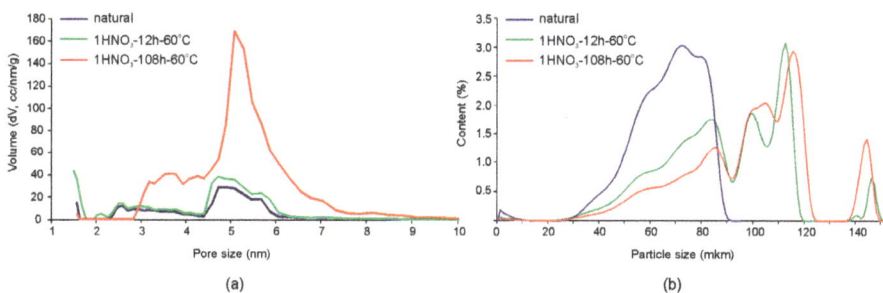

Figure 6. Changes of the porosity (**a**) and the particle size distribution (**b**) of Mt-T2 a sample due to exposure to 1 M HNO_3 at 60 °C.

Adsorption-desorption isotherms of N_2 are shown in Figure 6. As shown earlier, exposure to 0.25 and 0.5 M HCl for 7 days at room temperature results only in minor changes in the chemical composition (Table 2). The same tendency is observed in changes of N_2 adsorption-desorption isotherms for samples Mt-M and Mt-T1 (Figure 6). Mt-T1 sample was treated only at 0.5 M HCl concentrations and is not shown in Figure 7 since the changes in the adsorption-desorption rates are within the error margin of the method.

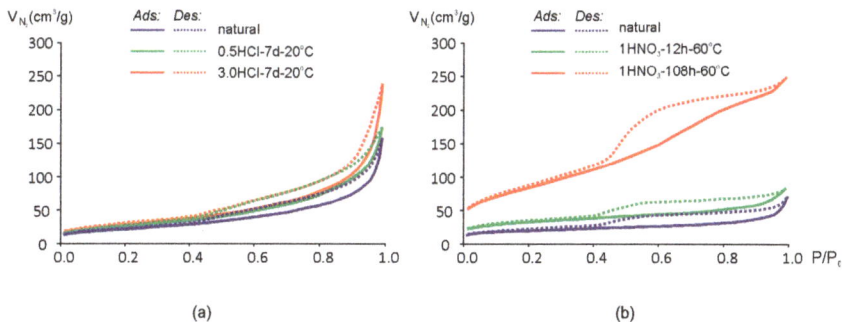

Figure 7. Isotherms of N_2 adsorption (Ads)/desorption (Des) of montmorillonites: (**a**) Mt-M; (**b**) Mt-T2.

Hysteresis loop and the shape of the isotherm is close to the type IVa [61,62], which means that the natural and acid-modified montmorillonites studied can be characterized as mesoporous adsorbents where monolayer-multilayer adsorption is followed by capillary condensation. There is a significant amount of the micropores and mesopores so N_2 adsorption/desorption isotherms cannot be described by one classical type only [61].

Hysteresis loops may have a different shape and, in the study of natural montmorillonite, can be attributed to the H3-H4 type [61]. Loops of this type correspond to the interaction between nonrigid lamellar particles. Smooth slope loops correspond to filling the micro- and mesopores.

Increase in steepness after the acid treatment may be explained by an increase of the pore volume and the appearance of macro-porosity. The specific hysteresis loop of the sample Mt-T2 isotherm after prolonged exposure to 1 M HNO_3 (Figure 7b) can be attributed to the type H2(b) [61]. This type of a desorption branch of the isotherm can be the result of a complex system of channel and pore formations and a partial blocking of them.

As a result of exposure to acid solutions, the hysteresis loop increases. In particular, it is noticeable after a prolonged influence of nitric acid at a high temperature (Figure 6b) when the hysteresis loop changes in a certain way.

The measured cation exchange capacity value (CEC) decreases from 86 meq/100 g in natural bentonite clay to 40 meq/100 g in bentonite samples with the longest time of the HNO_3 solutions treatment for 108 h (Table 1). Such a decrease of CEC under the treatment with inorganic acid solutions (e.g., HCl and H_2SO_4) was described by other researchers [12]. Cation exchange capacity of bentonite clays is connected not only to montmorillonite content but primarily, to such structural particularities as layer charge and its distribution among octahedral sheets. As a result, treatment with inorganic acid solutions leads to layer charge reduction due to leaching of the octahedral cations and the structural degradation; all of which leads to a decrease in the cation exchange capacity. In addition, amorphous silica appeared as a result of the destruction of tetrahedral sheets. The silica can settle on the surface of montmorillonite particles and reduce its exchange capacity.

A lot of attention in the work on acid treatment of natural bentonite clay was dedicated to the issues of activation—i.e., improving the properties of modified clays in comparison with natural clays [63,64]. An increase of specific surface area and pore volume is observed in the course of the conducted experiments (Table 1). Leaching of the cations from interlayer space and octahedral positions results in a modification of layer charge and particles in general, this in turn affects the interaction of the individual particles between each other. As a result, the destruction of large aggregates, the restructuring of smaller ones, the appearance of an uneven surface, and the appearance of micropores on the site of octahedral grid take place. All the processes described above result in an increase of the total pore size and surface area. However, opportunities for cation exchange and its capacity are reduced due to layer charge modification and protonation of interlayer space.

4. Conclusions

The study concludes that an interlayer modification occurs even at early stages of treatment with HCl and HNO_3 solutions. This modification involves partial substitution of interlayer cations, especially Ca and Mg, to oxonium and partial protonation of the interlayer space. As the least stable components, natural nano-sized smectites and the most defective phase are completely dissolved in natural montmorillonite samples after acidic treatment.

Further treatment of montmorillonite structure with inorganic acid solutions (with increasing concentration and exposure period) leads to further modification of its structure accompanied by intense leaching of cations from the octahedral positions and partial penetration of leached octahedral cations into interlayer space. These transformations lead to partial protonation of OH-groups, partial destruction of the octahedral sheets, and modification of the interaction between tetrahedral and octahedral sheets which changes layer charge and the nature of interaction between adjacent layers and partial amorphization of tetrahedral sheets. A decrease in the layer charge leads to a decrease in CEC.

As a result of structural transformation, specific surface area increases. Also, volume and pore diameters increase as well as micropores of a 3.5 nm diameter occur which is not typical for natural montmorillonites. As a consequence, particle size distribution changes. Due to the rise in micro- and mesoporosity, adsorption capacity of the acid-treated bentonite clays still remains high.

Modification of structural and adsorption characteristics with the acid treatment can be used to simulate behavior of the engineered barrier properties for repositories of radioactive and industrial wastes, especially in the case of dealing with liquid radioactive wastes.

Acknowledgments: This work was financially supported by the Russian Science Foundation (Project #16-17-10270). We thank Boris Pokid'ko for his help with analytical approach, and we also thank the "Bentonite" company for providing us with natural Bentonite clay from the Taganskoe deposit.

Author Contributions: Victoria V. Krupskaya prepared material, conceived and designed the experiments, and wrote the paper; Sergey V. Zakusin performed the experiments by HNO₃ treatment and measured samples by X-ray diffraction and CEC; Ekaterina A. Tyupina analyzed specific surface area and porosity; Olga V. Dorzhieva collected data by infrared spectroscopy; Anatoliy P. Zhukhlistov studied samples by electron microscopy; Maria N. Timofeeva performed the experiments by HCl treatment acid treatment; and Petr E. Belousov prepared <1 μm fractions and measured CEC.

Conflicts of Interest: The authors declare no conflict of interest.

References

1. Pusch, R.; Knutsson, S.; Al-Taie, L.; Mohammed, M.H. Optimal ways of disposal of highly radioactive waste. *Nat. Sci.* **2012**, *4*, 906–918. [CrossRef]
2. Sellin, P.; Leupin, O.X. The use of clay as an engineered barrier in radioactive-waste management—A review. *Clays Clay Miner.* **2013**, *61*, 477–498. [CrossRef]
3. Laverov, N.P.; Velichkin, V.I.; Omelianenko, B.I.; Yudincev, S.V.; Petrov, V.A.; Bichkova, A.V. Part 5: Changes of the environment and climate. In *Isolation of Spent Nuclear Materials: Geological and Geochemical Bases*; IGEM RAS, IFZ RAS: Moscow, Russia, 2008; p. 280. (In Russian)
4. Drits, V.A.; Choubar, C. *X-ray Diffraction by Disordered Lamelar Structure. X-ray Diffraction by Disordered Lamelar Structure*; Springer-Verlag: Berlin/Heidelberg, Germany, 1990; p. 371.
5. Moore, D.M.; Reynolds, R.C., Jr. *X-ray Diffraction and the Identification and Analysis of Clay Minerals*, 2nd ed.; Oxford University Press: Oxford, UK, 1997; p. 378.
6. Guggenheim, S.; Adams, J.M.; Bain, D.C.; Bergaya, F.; Brigatti, M.F.; Drits, V.A.; Formoso, M.L.L.; Gala, N.E.; Kogure, T.; Stanjek, H. Summary of recommendations of nomenclature committees. Relevant to clay mineralogy: Report of the Association Internationale Pour L'etude des Argiles (AIPEA) Nomenclature Committee for 2006. *Clays Clay Miner.* **2006**, *54*, 761–772. [CrossRef]
7. Brindley, G.W.; Brown, G. (Eds.) *Crystal Structures of Clay Minerals and Their X-ray Identification*; Mineralogical Society: London, UK, 1980.
8. Wilson, M.J. *Rock-Forming Minerals. Sheet Silikates: Clays Minerals*; The Geological Society: London, UK, 2013; p. 724.
9. Laverov, N.P.; Yudintsev, S.V.; Kochkin, B.T.; Malkovsky, V.I. The Russian strategy of using crystalline rock as a repository for nuclear waste. *Elements* **2016**, *12*, 253–256. [CrossRef]
10. Gupalo, T.A.; Kudinov, K.G.; Jardine, L.J.; Williams, J. Development of a Comprehensive Plan for Scientific Research, Exploration, and Design: Creation of an Underground Radioactive Waste Isolation Facility at the Nizhnekansky Rock Massif. In Proceedings of the Waste Management 2005 Symposium, Tucson, AZ, USA, 27 February–3 March 2005.
11. Komadel, P. Chemically modified smectites. *Clay Miner.* **2003**, *38*, 127–138. [CrossRef]
12. Tomić, Z.P.; Antić Mladenović, S.B.; Babić, B.M.; Poharc Logar, V.A.; Đorđević, A.R.; Cupać, S.B. Modification of smectite structure by sulfuric acid and characteristics of the modified smectite. *J. Agric. Sci.* **2011**, *56*, 25–35.
13. Okada, K.; Arimitsu, N.; Kameshima, Y.; Nakajima, A.; MacKenzie, K.J.D. Solid acidity of 2:1 type clay minerals activated by selective leaching. *Appl. Clay Sci.* **2006**, *31*, 185–193. [CrossRef]
14. Kumar, P.; Jasra, R.V.; Bhat, T.S.G. Evolution of Porosity and Surface Acidity in Montmorillonite Clay on Acid Activation. *Ind. Eng. Chem. Res.* **1995**, *34*, 1440–1448. [CrossRef]
15. Komadel, P.; Madejova, J. Chapter 7.1: Acid activation of clay minerals. In *Handbook of Clay Science*; Developments in Clay Science; Bergaya, F., Theng, B.K.G., Lagaly, G., Eds.; Elsevier: Amsterdam, The Netherlands, 2006; Volume 1, p. 263287.
16. Carrado, K.A.; Komadel, P. Acid activation of bentonites and polymer-clay nanocompo-sites. *Elements* **2009**, *5*, 111–116. [CrossRef]
17. Dubikova, M.; Cambier, P.; Sucha, V.; Caplovicova, M. Experimental soil acidification. *Appl. Geochem.* **2002**, *17*, 245–257. [CrossRef]
18. Pagano, T.; Sergio, M.; Glisenti, L.; Diano, W.; Grompone, M.A. Use of pillared montmorillonites to eliminate chlorophyll from rice bran oil. *Ing. Quim.* **2001**, *19*, 11–19.

19. Tyupina, E.A.; Magomedbekov, E.P.; Tuchkova, A.I.; Timerkaev, V.B. The sorption refinement of liquid organic radioactive waste for Cs-137. *Adv. Mater. Spec. Issue* **2010**, *8*, 329–333. (In Russian).

20. Vicente, M.A.; Suarez Barrios, M.; Lopez Gonzalez, J.D.; Banares Munoz, M.A. Characterization, surface area, and porosity analyses of the solids obtained by acid leaching of a saponite. *Langmuir* **1996**, *12*, 566–572. [CrossRef]

21. Komadel, P. Structure and chemical characteristics of modified clays. In *Natural Microporous Materials in Environmental Technology*; Misealides, P., Macasek, F., Pinnavaia, T.J., Colella, C., Eds.; Kluwer: Alphen aan den Rijn, The Netherlands, 1999; pp. 3–18.

22. Tkac, I.; Komadel, P.; Muller, D. Acid-treated Montmorillonites—A Study by 29Si and 27Al MAS NMR. *Clay Miner.* **1994**, *29*, 11–19. [CrossRef]

23. He, H.; Guo, J.; Xie, X.; Lin, H.; Li, L. A microstructural study of acid-activated montmorillonite from Choushan, China. *Clay Miner.* **2002**, *37*, 337–344. [CrossRef]

24. Timofeeva, M.N.; Panchenko, V.N.; Gil, A.; Zakusin, S.V.; Krupskaya, V.V.; Volcho, K.P.; Vicente, M.A. Effect of structure and acidity of acid modified clay materials on synthesis of octahydro-2H-chromen-4-ol from vanillin and isopulegol. *Catal. Commun.* **2015**, *69*, 234–238. [CrossRef]

25. Novikova, L.; Belchinskaya, L.; Krupskaya, V.; Roessner, F.; Zhabin, A. Effect of acid and alkaline treatment on physical-chemical properties of surface of natural glauconite. *Sorpt. Chromatogr. Process.* **2015**, *15*, 730–740.

26. Timofeeva, M.N.; Volcho, K.P.; Mikhalchenko, O.S.; Panchenko, V.N.; Krupskaya, V.V.; Tsybulya, S.V.; Gil, A.; Vicente, M.A.; Salakhutdinov, N.F. Synthesis of octahydro-2H-chromen-4-ol from vanillin and isopulegol over acid modified montmorillonite clays: Effect of acidity on the Prins cyclization. *J. Mol. Catal. A Chem.* **2015**, *398*, 26–34. [CrossRef]

27. Franco, F.; Pozo, M.; Cecilia, J.A.; Benítez-Guerrero, M.; Lorente, M. Effectiveness of microwave assisted acid treatment on dioctahedral and trioctahedral smectites. The influence of octahedral composition. *Appl. Clay Sci.* **2016**, *120*, 70–80. [CrossRef]

28. Adams, J.M. Synthetic organic chemistry using pillared, cation-exchanged and acid- treated montmorillonite catalysts—A review. *Appl. Clay Sci.* **1987**, *2*, 309–342. [CrossRef]

29. Brown, D.R. Review: Clays as catalyst and reagent support. *Geol. Carpath. Ser. Clays* **1994**, *45*, 45–56. [CrossRef]

30. Bovey, J.; Jones, W. Characterization of Al-pillared acid-activated clay catalysts. *J. Mater. Chem.* **1995**, *5*, 2027–2035. [CrossRef]

31. Timofeeva, M.N.; Panchenko, V.N.; Volcho, K.P.; Zakusin, S.V.; Krupskaya, V.V.; Gil, A.; Mikhalchenko, O.S.; Vicente, M.A. Effect of acid modification of kaolin and metakaolin on Brønsted acidity and catalytic properties in the synthesis of octahydro-2H-chromen-4-ol from vanillin and isopulegol. *J. Mol. Catal. A Chem.* **2016**, *414*, 160–166. [CrossRef]

32. Timofeeva, M.N.; Panchenko, V.N.; Krupskaya, V.V.; Gil, A.; Vicente, M.A. Effect of nitric acid modification of montmorillonite clay on synthesis of sotketal from glycerol and acetone. *Catal. Commun.* **2017**, *90*, 65–69. [CrossRef]

33. Tokarev, I.V.; Rumynin, V.G.; Zubkov, A.A.; Pozdnyakov, S.P.; Polyakov, V.A.; Kuznetsov, V.Y.U. Assessment of the long-term safety of radioactive waste disposal: 1. Paleoreconstruction of groundwater formation conditions. *Water Resour.* **2009**, *36*, 206–213. [CrossRef]

34. Zubkov, A.A.; Balakhonov, B.G.; Sukhorukov, V.A.; Noskov, M.D.; Kessler, A.G.; Zhiganov, A.N.; Zakharova, E.V.; Darskaya, E.N.; Egorov, G.F.; Istomin, A.D. Radionuclide distribution in a sandstone injection zone in the course of acidic liquid radioactive waste disposal. *Dev. Water Sci.* **2005**, *52*, 491–500.

35. Rybalchenko, A.; Pimenov, M.; Kostin, P. Injection Disposal of Hazardous and Industrial Wastes, Scientific and Engineering Aspects. In *Deep Injection Disposal of Liquid Radioactive Waste in Russia*; Academic Press: New York, NY, USA, 1998; p. 780.

36. Rybalchenko, A.I.; Pimenov, M.K.; Kurochkin, V.M.; Kamnev, E.N.; Korotkevich, V.M.; Zubkov, A.A.; Khafizov, R.R. Deep Injection Disposal of Liquid Radioactive Waste in Russia. *Dev. Water Sci.* **2005**, *52*, 13–19.

37. Utkin, S.S.; Linge, I.I. Decommissioning strategy for liquid low-level radioactive waste surface storage water reservoir. *J. Environ. Radioact.* **2016**. [CrossRef] [PubMed]

38. Malkovsky, V.I.; Dikov, Y.P.; Asadulin, E.E.; Krupskaya, V.V. Influence of host rocks on composition of colloid particles in groundwater at the Karachai Lake site. *Clay Miner.* **2012**, *47*, 391–400. [CrossRef]

39. Post, J.E.; Bish, D.L. Rietveld refinement of crystal structures using powder X-ray diffraction data. *Rev. Mineral. Geochem.* **1989**, *20*, 277–308.

40. Doebelin, N.; Kleeberg, R. Profex: A graphical user interface for the Rietveld refinement program BGMN. *J. Appl. Crystallogr.* **2015**, *48*, 1573–1580. [CrossRef] [PubMed]

41. Czímerová, A.; Bujdák, J.; Dohrmann, R. Traditional and novel methods for estimating the layer charge of smectites. *Appl. Clay Sci.* **2006**, *34*, 2–13. [CrossRef]

42. Zakusin, S.V.; Krupskaya, V.; Dorzhieva, O.V.; Zhuhlistov, A.P.; Tyupina, E.A. Modification of the adsorption properties. *Sorpt. Chromatogr. Processes* **2015**, *15*, 280–289.

43. Fineevich, V.P.; Allert, N.A.; Karpova, T.R.; Dupliakin, V.K. Composite nanomaterials based on acid-activated montmorillonite. *Russ. Chem. J.* **2007**, *4*, 69–74. (In Russian)

44. Tyagi, B.; Chudasama, C.D.; Jasra, R.V. Determination of structural modification in acid activated montmorillonite clay by FT-IR spectroscopy. *Spectrochim. Acta Part A* **2006**, *64*, 273–278. [CrossRef] [PubMed]

45. Rhodes, C.N.; Brown, D.R. Catalytic activity of acid-treated montmorillonite in polar and nonpolar reaction media. *Catal. Lett.* **1994**, *24*, 285–291. [CrossRef]

46. Shlikov, V.G. *X-ray Analysis of Mineral Composition of Fine-Grained Soil*; GEOS: Moscow, Russia, 2006; p. 175. (In Russian)

47. Madejova, J.; Komadel, P. Baseline studies of The Clay Minerals Society Source Clays: Infrared methods. *Clays Clay Miner.* **2001**, *49*, 410–432. [CrossRef]

48. Russell, J.D.; Fraser, A.R. *Clay Mineralogy: Spectroscopic and Chemical Determinative Methods*; Wilson, M.J., Ed.; Chapman & Hall: London, UK, 1996; pp. 11–67.

49. Pentrák, M.; Czímerová, A.; Madejová, J.; Komadel, P. Changes in layer charge of clay minerals upon acid treatment as obtained from their interactions with methylene blue. *Appl. Clay Sci.* **2012**, *55*, 100–107. [CrossRef]

50. Komadel, P. Acid activated clays: Materials in continuous demand. *Appl. Clay Sci.* **2016**, *131*, 84–99. [CrossRef]

51. Dong, H.; Peacor, D.R. TEM observations of coherent stacking relations in smectite, i/s and illite of shales: Evidence for Macewan crystallites and dominance of 2M1 polytypism. *Clays Clay Miner.* **1996**, *44*, 257–275. [CrossRef]

52. Osipov, V.I.; Sokolov, V.N. *Clays and Its Properties. Composition, Structure and Properties Formation*; GEOS: Moscow, Russia, 2013; p. 576. (In Russian)

53. Simc, V.; Uhlík, P. Crystallite size distribution of clay minerals from selected Serbian clay deposits. *Geoloski Anali Balkanskoga Poluostrva* **2006**, *67*, 109–116.

54. Kotarba, M.; Srodon, J. Diagenetic evolution of crystallite thickness distribution of illitic material in Carpathian shales, studied by the Bertaut-Warren-Averbach XRD method (MudMaster computer program). *Clay Miner.* **2000**, *35*, 383–391. [CrossRef]

55. Mystkowski, K.; Srodon, J. Mean thickness and thickness distribution of smectite crystallites. *Clay Miner.* **2000**, *35*, 545–557. [CrossRef]

56. Trofimov, V.T.; Korolev, V.A.; Voznesensky, E.A.; Golodovskaya, G.A.; Vasil'chuk, Y.K.; Ziangirov, R.S. *Soil Science*, 6th ed.; MSU: Moscow, Russia, 2005; p. 1024. (In Russian)

57. Rouquerolt, J.; Avnir, D.; Fairbridge, C.W.; Everett, D.H.; Haynes, J.H.; Pernicone, N.; Ramsay, J.D.F.; Sing, K.S.W.; Unger, K.K. Recommendations for the characterization of porous solids. *Pure Appl. Chem.* **1994**, *66*, 1739–1758.

58. Karnauhov, A.P. *Adsorption. Texture of Dispersive and Porous Materials*; Nauka: Novosibirsk, Russia, 1999; p. 470. (In Russian)

59. IUPAC. Manual of Symbols and Terminology. *Pure Appl. Chem.* **1972**, *31*, 577.

60. Churakov, S. Mobility of Na and Cs on Montmorillonite Surface under Partially Saturated Conditions. *Environ. Sci. Technol.* **2013**, *47*, 9816–9823. [CrossRef] [PubMed]

61. Thommes, M.; Kaneko, K.; Neimark, A.V.; Olivier, J.P.; Rodriguez-Reinoso, F.; Rouquerol, J.; Sing, K.S.W. Physisorption of gases, with special reference to the evaluation of surface area and pore size distribution (IUPAC Technical Report). *Pure Appl. Chem.* **2015**, *87*, 1051–1069. [CrossRef]

62. Novikova, L.; Ayrault, P.; Fontaine, C.; Chatel, G.; Jérôme, F.; Belchinskaya, L. Effect of low frequency ultrasound on the surface properties of natural aluminosilicates. *Ultrason. Sonochem.* **2016**, *31*, 598–609. [CrossRef] [PubMed]

63. Kheok, S.C.; Lim, E.E. Mechanism of palm oil bleaching by montmorillonites clay activated at various acid concentrations. *J. Am. Oil Chem. Soc.* **1982**, *59*, 129–131. [CrossRef]
64. Morgan, D.A.; Shaw, D.B.; Sidebottom, T.C.; Soon, T.C.; Taylor, R.S. The function of bleaching earth in the processing of palm, palm kernel and coconut oils. *J. Am. Oil Chem. Soc.* **1985**, *62*, 292–299. [CrossRef]

MDPI AG

St. Alban-Anlage 66

4052 Basel, Switzerland

Tel. +41 61 683 77 34

Fax +41 61 302 89 18

http://www.mdpi.com

Minerals Editorial Office

E-mail: minerals@mdpi.com

http://www.mdpi.com/journal/minerals